Shelly Chauhan, a chartered p<s>s</s> has coached and trained a large n<s>...</s> world over the past eighteen years on issues such as resilience, stress management and emotional intelligence. In sessions with clients, the conversation regularly turned to parenting as they sought to discover how they could manage their stress levels to be better parents. Having two young children herself, Shelly was only too aware of the challenges of parenting and how parenting guilt and worry could affect other areas of one's life. Over the years she came to realise that at the heart of many of these clients' difficulties was their struggle to notice and regulate their emotional states, and how deeply this affected their ability to connect with, influence and feel genuine empathy for other people, including their children.

Based on years of meticulous research into the neuroscience of emotional regulation, stress and human connection, Shelly was inspired to translate this complex research into a methodology combining brain science with practical exercises to help clients enhance their capacity for emotional regulation, wellbeing and connection. Feedback from clients highlights the power of this approach to transform parenting and increase resilience, empathy and general wellbeing both in children and parents. Shelly has a deep sense of compassion towards children. She is committed to helping parents understand how powerful an impact loving connection has on the growing brains of their children and how the need to be loved, understood and accepted is hardwired into us all. Shelly lives in Surrey, England, with her husband and two children, aged ten and seven.

Heartfelt Parenting

Shelly Chauhan

A HOW TO BOOK

ROBINSON

ROBINSON

First published in Great Britain in 2020
by Robinson

10 9 8 7 6 5 4 3 2 1

A CIP catalogue record for this book
is available from the British Library.

ISBN: 978-1-47214-122-4

Typeset in Sentinel by
Initial Typesetting Services, Edinburgh
Printed and bound in Great Britain by
Clays Ltd, Elcograf S.p.A.

Papers used by Robinson are from well-
managed forests and other responsible
sources.

Robinson
An imprint of
Little, Brown Book Group
Carmelite House
50 Victoria Embankment
London EC4Y 0DZ

An Hachette UK Company
www.hachette.co.uk

www.littlebrown.co.uk

....

How To Books are published by Robinson,
an imprint of Little, Brown Book Group. We
welcome proposals from authors who have
first-hand experience of their subjects.
Please set out the aims of your book, its target
market and its suggested contents in an email
to howto@littlebrown.co.uk

....

To my beautiful, warm-hearted children without whom
I would never have learnt all that I have about what
love really is. My gratitude is endless.

Out beyond ideas of wrongdoing and rightdoing,
there is a field. I'll meet you there.
When the soul lies down in that grass,
the world is too full to talk about.
Ideas, language, even the phrase 'each other'
doesn't make any sense.

– Rumi

CONTENTS

Introduction

We, as parents, have access to an ever-growing body of information about parenting and psychology, and we're becoming increasingly aware of how the way in which we interact with our children can shape and mould their behaviour. But is all this information helping us focus on what our children really need from us in order to develop healthy brains, hearts and relationships?

We often find ourselves going through the motions of parenting, based on what we have read and our worries about what might happen to our children if we don't follow all that advice. When my first child was born, I recall reading books on how parenting shapes a child's brain for life and although I thought I understood them at the time, and I tried hard to follow the advice in those books, I now know that I was 'doing' good parenting rather than allowing it to emerge naturally from a state of heartfelt connection. Through years of study and personal practice both as a psychologist and a mother, I slowly learnt how to be calm, warm and connected with my children, and that state of 'being' has had a transformational impact on our family life. I finally understand how and why parenting can seem so tough at times, and what needs to happen for it to flow more easily and intuitively.

Connection is not something you can read about and do. It is an emotional state, based on micro-moments of emotional synchrony between two people, that generates a visceral feeling of being 'in synch' at that point in time. It facilitates understanding and empathy between people and is as critical to the development of a healthy brain and mind as oxygen is for our lungs. Connection is at the heart of wellbeing because the very same biological system that promotes connection also enables us to feel calm, regulate our emotions, rest and restore our bodily organs and feel open to our experiences. Not only that but children need this kind of connection to grow brains that are 'joined up' in a way that promotes emotional resilience, something that is increasingly necessary in an age in which anxiety, stress and depression are seemingly on the rise, in both children and adults.

We have entered a new era in terms of understanding the neuroscience of how connection influences and shapes us as human beings. Nowhere is this more relevant and profound than in relation to children and what they really need from their parents. We know that humans have a dedicated system in our brains and bodies to help us to assess whether we are safe or must prepare to defend ourselves from a threat. This is based on millions of years of evolution through which we have had to maintain a very high level of vigilance to ensure our survival. If we could hear this threat-safety detection system speak, it would be asking and answering the questions, 'Am I safe?', 'Are you with me?'. This occurs in less than a second and takes place in a perpetual loop outside of your conscious awareness.

It is only when the answer to those questions is 'Yes' that we can relax, feel comfortable in our bodies and minds, and interact with other people in a way that sustains connection and warmth. Feelings of safety emerge when we are understood and accepted. It is the equivalent of having someone tell us, 'I see what you are feeling right now', 'I'm here for you' and 'I accept you as you are'. Note that this is a non-verbal process and requires very little talking, if any. It stems from a very different brain system to the anxious, judgemental, critical or angry interactions that so often compromise the parenting relationship. When you connect in a calm, reciprocal way with someone, you produce oxytocin, which is a powerful hormone that can promote bonding and social behaviour but also reduces our reactivity to stress and anxiety.

Good parenting rests on the process of attunement, a natural, empathic understanding of the emotional states of another person. Emotions begin as tiny shifts in our inner bodily state – our heart rate, breathing, muscle tension and level of chemicals (neurotransmitters and hormones, for example) may all go up or down in different patterns that are unique to each category of emotion. Anger, for example, will lead to a rise in your heart rate whereas sadness will lower it. When you are able to tune in with what is happening in your body and regulate it, you can remain calm yet alert when you are faced with something that you find hard to handle. When you are unable to regulate your inner emotional state, you can flip into either a high level of emotional activation that makes you want to vent the emotion or try to avoid it, or you might shut down. Stress, anger, agitation, frustration, excitability, anxiety, disgust and contempt all represent states of high activation, which lead to dwindling stores of empathy and connection. To connect in a heartfelt way, you need to be regulated, and to regulate emotions you need to be able to feel emotional states in your own body. It is only when you can do these things that you become capable of regulating the emotions of your children.

This matters for two reasons. Firstly, our children cannot regulate their own emotions in a sophisticated manner until they are much older, sometimes not until they are in their mid-twenties, which means that they are easily swayed by strong emotional currents that toss them in all directions, seldom the one we want them to go in. Until they develop those vital parts of the brain that underpin emotional regulation, self-control, self-awareness and decision-making, they must rely on us to be a 'proxy' brain for them, mirroring to them what they are feeling and helping them back to states of calm.

Secondly, because our children rely on us for their sense of inner safety and also for their sense of who they are, they become vulnerable and stressed when we are not available to notice and care about how they are feeling, and we are unable to calm them when they need it. They come into the world with all the right biological equipment to be able to 'read' us and infer (non-verbally, in terms of 'threat' versus 'safety') where they stand with us moment by moment. Are we tuned in? Do we care? Are we available to them? Will we notice what they need? Will they survive? This need for connection is hardwired into us. When we bear in mind that emotions involve real physiological changes to heart rate, chemicals and muscle tension, all of which can feel uncomfortable, and that our children have no 'off' button for their big emotions – like having blaring music played on a sound system with no volume control – it becomes so much easier to understand why we need to create the conditions for connection to emerge and thrive in our relationships with them.

More importantly, our children sense our emotional state, from tiny micro-movements of the muscles of our eyes, and minuscule shifts in the tone of our voice, outside of conscious awareness. When they do, their inner emotional states often synchronise with ours because these states can be contagious, leading to cascades of negative emotions between parents and children, creating stress and annoyance. Parenting really is a heartfelt process because our hearts are connected, via the vagus nerve, to the facial muscles we use to express emotions, meaning that when we have small shifts in emotion that register in our voices and on our faces, it can quicken or slow down the heart rates of our children. This is one of the mechanisms behind how we come to share emotions with each other.

When you have a strong foundation of connection with your children, parenting starts to feel easier, calmer and more joyful. Rather than feeling as though you are depleting your inner resources with your children, you start to feel nourished and content when you are around them. But surely, you might be thinking, connection just happens naturally and doesn't require a book delving into the underlying brain science of it? Some of us may well find this to be the case but I'm convinced that we are more at risk of disconnection than ever before, and, what's more, I have some wonderful science to share with you that backs this up. The way in which we use our brains is changing over time, with an increasing emphasis on brain systems that promote doing, striving, achieving, judging, controlling and thinking, and less of an ability to harness those brain regions that promote bodily awareness, acceptance, stillness, being present in the moment, and emotional connection. This has a very real impact on our relationships with our children, which in turn affects their sense of themselves; how they learn to regulate emotions; how they deal with adversity; how they relate to other people, and ultimately, I believe, how lovable they come to believe they are. When we are stressed and busy, it is much harder, if not impossible, to access our natural inbuilt caring system from which connection emerges and serves its wonderful job of promoting emotional safety, health and wellbeing.

This book is different because rather than focus on how you want to influence your children and what you want for them, it focuses on you and how, in learning to give your children what their brains and bodies long for – acceptance, understanding and warmth – you will automatically build the foundations for those very outcomes you seek for your children and your family life. And for those of you who might need a little more than an emotional eulogy for connection, I'm going to back this up with some influential neuroscience based on the work of experts who have spent years studying emotion, interpersonal neurobiology (how relationships shape our brains and vice versa) and the emotional brain. But, above all, I'm writing this as a parent because my experience of using the knowledge and methods presented in this book has had a momentous impact on how parenting feels to me and my children at a deep visceral level.

To get the most from this book, I would advise you to read it slowly and give yourself time to digest each chapter before moving on to the next. It has taken me several years to comprehend fully the impact of the science I'm going to share with you, and the more I reflect on it the more I benefit from it. But the most important point I want to make in this book is that connection is a process involving the brain, body and mind, and it can only surface when your nervous system is in a regulated and relaxed state. This means that reading alone won't be sufficient for this type of parenting to flow. You will need to learn to regulate your emotions through calming your nervous system, which involves breathing exercises that will require regular practice and commitment. As a corporate psychologist who coaches executives on emotional regulation and stress management, I know several of my clients have been sceptical of these breathing exercises at first. But once they understand the neuroscience these methods are based on, and they experience the benefits of regular practice, many of them report positive changes not just at work, but at home, in their sleep patterns, levels of wellbeing, and most often with their children.

CHAPTER ONE

What is parenting really about?

Parenting, at its core, is about developing and maintaining a unique and special relationship that enables children to feel safe, loved, supported and accepted whilst they grow into themselves and find their place in the world. Many parents, when asked what they most want for their children, say something involving the words, 'happy' and/or 'successful'. We want our children to have good lives, though we may have different notions of what a good life entails. Some of us may assume that providing opportunities for our children to polish their academic achievements and go to a good university might secure them a future of happiness and stability. Others might think protecting their children from difficulties, either physical or emotional, will cocoon them against unhappiness. Sometimes we assume children need things we never achieved or experienced ourselves, or that they must be protected from the distressing experiences we ourselves had when growing up.

But do we really know what contributes to a good and fulfilling life? Are we parents, for all our efforts and intentions, really giving our children the best chances of acquiring what we most want for them? And in pursuing some of the outcomes we believe most beneficial to our children, are we getting lost in the peripheral noise and bypassing the most important contribution that parenting can make to their lives?

What if I told you that this book could help you find a scientifically proven, neuroscience-backed, natural remedy for children that makes them more likely to grow into resilient, confident and happy individuals? This powerful natural remedy lowers levels of stress and anxiety, ramps up their immune systems, and offers them a buffer against mental illnesses. It helps prevent tantrums, calms emotional outbursts and miraculously makes your children want to please you. It has no clinical side effects, can be administered anytime and has no dose limitations. Even if you think this sounds too good to be true, you probably also want to find and stockpile this substance in industrial quantities, no matter the cost! Well, the good news is that you are the laboratory that produces this remarkable substance and, depending on early life experiences, most of us can produce it at will; the bad news is that it can only be made when certain conditions are in place and unfortunately the habits of modern life make these increasingly rare.

CONNECTION IS AT THE HEART OF PARENTING

The remedy I'm referring to here is 'connection'; a deep, non-verbal sense of closeness between two human beings that allows them to tune in with each other. This type of connection is defined by a level of 'in-the-moment' synchrony between parts of your brain and the brain of your child; it is underpinned by the activation of a certain branch of your nervous system, and in particular, a wandering nerve called the vagus nerve, which influences your breathing, heart rate, tone of voice, facial muscles and level of eye contact. It causes the release of powerful chemicals that cascade through your body allowing you to relax and bond with each other, feeling a mutual sense of trust and openness.

This form of connection is about micro-moments of shared emotion, what Barbara Fredrickson[1] calls 'Positivity Resonance', a spiral of mirrored positive feeling that boosts immunity, improves cardiac health, increases resilience and actually broadens the mind. This level of shared emotion between the brains and bodies of two or more people, breaks down a sense of 'I' and creates a sense of 'we', bringing about a momentary feeling of fusing together. Frequent shared experiences of these moments of resonance and synchrony create the relationships children crave with us. Feeling these moments of connection is very different to knowing you love your child; and the important thing for parents to know is that this feeling part of love really matters because it shapes your child's brain, emotional habits and later relationships.

Human brains are hardwired for this type of connection and it is as essential to the growth of a healthy brain as food and water are to the body. Connection at this level has a wide-ranging impact on many markers of physical and emotional wellbeing, right down to how long your child may live. Early scientific findings, based on the work of Barbara Fredrickson, reveal that it may positively alter the way DNA is expressed in the genes, so it is not just a source of emotional comfort to you and your child; it physically nourishes and changes them at a cellular level. This is a powerful reminder of the impact we have on each other and how deeply entangled emotional connection is with physical health. Convinced as we may be of its value, being connected is not as easy as it sounds because to feel this kind of connection, we need to cultivate a certain state in our bodies and minds, something that seems increasingly difficult in this era of accelerated change and advancement. This book will show you why connection affects the wellbeing of our children, why it is slowly being eroded and, more importantly, what we need to do to allow it to flourish.

WHAT'S GETTING IN THE WAY OF CONNECTION?

Parenting in this day and age seems imbued with a sense of near oppressive responsibility and, at times, even anxiety. We live in a time when parenting is no longer what we do simply by bringing children into the world and providing for their needs; it is no longer an accidental by-product of lived experience with our children, as it was for our parents. Parenting now seems to be a skill to think about, much like any other, something to be studied, analysed and continually improved using a bewildering and often conflicting plethora of tools and techniques.

It is not enough to allow our children to evolve through chance and circumstance but rather this evolution is something we must influence and control. Just remembering all the parenting advice we're exposed to is hard enough, let alone trying to implement any of it when we're tired, stressed and in the grip of strong emotions. This makes us worry that we're somehow getting it wrong.

Responsibility without control – a recipe for stress

We now know so much more about how children develop than our parents and grandparents ever did, and this information, although enlightening, also leads to worry and guilt. Increased knowledge places a pressure on us to ensure our kids turn out well but paradoxically we have less control over them than ever before, mostly because the more we learn about parenting, the more we try to parent using methods of empowerment rather than fear and threat. But the fear of punishment has traditionally been the way parents and teachers got children to toe the line and behave themselves, so we're in a situation where we feel responsible for how they turn out yet we feel we have less direct control over them. This is a recipe for stress if ever there was one.

The challenge of balancing authority with warmth

Parenting without shaming, blaming and threat-based behaviour management is a worthy commitment but also, for many parents, fraught with uncertainty and doubt. We simply don't know enough to have a lasting confidence in what we're doing with our children and so we swing between being lenient and being authoritarian, feeling guilty and worried at both ends of this spectrum. How will we get them to listen without making them afraid of what will happen when they don't? How will they come to respect us if they don't fear the consequences of not doing so? Won't they just go off the rails if we're too soft? But on the other hand, if we push too hard they might end up with anxiety or depression. After all, we're regularly told we're in the midst of an epidemic and these mental conditions are on the rise in young children and teenagers. Or worse, maybe they won't love us enough if we're too tough on them. Many of us have an intuitive understanding that pushing too hard can rupture that precious sense of connection that is so vital to parenting.

It's hard to figure out how to find a balance between nurturing your child and wielding authority. But that's exactly what good parenting is about: high warmth (connection) combined with high authority (having high expectations and enforcing them without frightening, blaming or shaming). As I'll elaborate later, connection and warmth must be the foundation upon which discipline and boundaries are layered. It is only when the connection is firmly in place that wielding authority won't interfere with some of the longer-term goals of parenting such as building resilience, optimism and self-acceptance in your child.

Are we parents increasingly anxious and worried?

Going back to what obscures our ability to connect with our children, there's another obstacle on the rise: we have too much choice and control over their lives, which brings with it increased anxiety to maximise opportunities lest your children end up missing out or unhappy. The combination of these anxieties may be a factor in the rise of so-called 'helicopter parenting' and other forms of parenting that focus on over-protection through parental involvement. Parenting seems increasingly driven by guilt and fear: fear of what might go wrong, of our children having to experience emotional discomfort, or of opportunities missed; and guilt that we may be the cause of it.

Many parents also get lost in the pull of the 'ego' and unknowingly make parenting just a little too much about themselves. They inadvertently impose their insecurities and emotions on their children without an awareness that they are doing so, often to the detriment of the child's sense of self. Some worry about social approval; they see their children's actions and traits as though they were being judged by an imaginary audience, and parenting becomes about avoiding shame, embarrassment, or scrutiny. It is hard to remember that our children are not an extension of ourselves, to be used to bolster or protect our levels of comfort or self-esteem; they are not here to be judged by whether they increase or reduce our social status and positive perception in the eyes of others. They are unique individuals with whom we must deeply connect but also differentiate ourselves from.

Judging our children through the eyes of others

Thanks to the rise of pop psychology and the relentless culture of self-improvement based on short-term behaviour management, parenting has become quite focused on controlling 'in-the-moment' behaviour, like tantrums and emotional outbursts, because we either can't tolerate how that behaviour feels to us or we can't handle how it looks to others. It's easy to forget that just because you know there are techniques parents can use to help regulate or control a child's behaviour doesn't mean children can or should behave perfectly all the time. It is also hard not to compare your children with their peers, either real or the vast virtual peer group created via the internet.

Just as parents can't be, and don't need to be, balanced and compassionate at every moment, our children mustn't be scrutinised and judged every time they step out of line. This level of scrutiny is not compassionate or healthy; it activates a brain system that can easily spiral out of control, leading to rigidity, anxiety and depression. It is no surprise that social media platforms that rely on approval-seeking and social judgement may damage the mental health of the children who overuse them. Your child's behaviour in the moment is not an inexorable indication of who your child will be in fifteen years' time. We must focus more on the long term and give our children a chance to grow into themselves without the fear of failing to reach unrealistic standards. I'm not saying the short term doesn't matter at all or that you should not pay attention to what typical children of the same age as yours might or might not do; I'm just saying there is another more vital area of parenting that we are neglecting to the detriment of our own and our children's emotional wellbeing.

Perhaps we're overcomplicating parenting and losing perspective

Of course, we're thinking about all this because we know much more about the influence we might have on our children as parents. This new-found parenting knowledge can be a blessing, an enlightening sea change in the world of child-rearing, but it can also lead to a tsunami of parental worry and confusion. Parenting in this day and age is like being given a 1000-piece jigsaw puzzle, tasked with having to piece it together perfectly, except it has no box and no pictures, so we have no idea what we are building. We take guesses and stabs at it but often with little clarity of what we are aiming for in the end. We get lost in the complex list of recommendations, suggestions and imperatives about what we should and shouldn't do with our children.

We're overcomplicating parenting because we're focusing too much on outcomes and not enough on the process. Focusing too much on your child's emotional and behavioural outcomes or achievements is sometimes counterproductive and often a recipe for anxiety. After all, no matter how many opportunities you throw their way now, there can be no certainty that your child will be happy, comfortable or successful later in life; and no matter what you do when they are young, you will never be able to control all the variables that life will fling at them once they are older. Sure, you can try to beat the odds by protecting them, pushing them, sheltering them or exposing them to things. But even if you do, they are going to have experiences that result in feelings of unhappiness, rejection, frustration, and sometimes even failure. These experiences don't have to scare you or them if you accept that the uncomfortable emotions that accompany such experiences are a natural and essential ingredient for building character and resilience.

We don't have to protect them from emotional discomfort

Emotional discomfort is only something to be afraid of if you and your children struggle to regulate (feel and calm) difficult emotions. As a parent you naturally feel responsible for protecting your children from bad experiences but you don't need to protect them from every experience that makes them feel bad. What matters so much more to their long-term development is, firstly, how you help them to regulate their emotional responses to difficult experiences and secondly, how you help them to make sense of the things that happen in their lives. This is the real crux of resilience.

Whilst bolstering their tolerance of difficult emotions, we also do want to nurture, in balanced ways, the capacity for positive emotions such as joy, curiosity, contentment, gratitude, love and so on. These are of a different nature to excitement and pleasure-seeking and tend to emerge from a state of slowing down and savouring small daily experiences rather than creating big special moments, though those also have their place occasionally.

ARE WE FOCUSING ON THINGS THAT REALLY MATTER IN DEVELOPMENTAL TERMS?

The alternative is to focus less on outcomes and more on the process of parenting. Rather than trying so hard to give them opportunities for success and pleasure, to remove obstacles and sources of discomfort to them, why not also help them to develop the emotional and mental traits that enable them to cope well with whatever life throws their way? This is a very different approach to, for example, sending them to every after-school activity going in the hope that greater exposure will somehow give them the edge over competitors later in life. Or in the hope that they will never accuse you of not having provided them with opportunities to exercise their interests and develop their skills. Or simply because they want to and you struggle to deny them things they want.

This approach is about paying attention to values and strengths that we know contribute to a good life. What I'm advocating is that we help them build qualities that we know have an impact on important life outcomes. Traits and abilities such as resilience, emotional regulation, persistence, attention and self-awareness will not guarantee them success, but certainly augment their chances of attaining it whilst protecting their mental and emotional wellbeing in the process. Some of these are essential components of what is called 'Emotional Intelligence'.

WHAT IS EMOTIONAL INTELLIGENCE AND WHY DOES IT MATTER?

In the 1990s, Daniel Goleman[2] popularised the notion of Emotional Intelligence (EQ) and since then, a growing body of research has identified the numerous benefits it may yield. Emotional Intelligence can be defined as the ability to recognise, understand and manage one's own and others' emotional reactions in ways that are appropriate to the situation in hand. Your child's level of Emotional Intelligence will play a part in how he responds to challenges, complexity and relationships later in life. At work, for example, we know that a certain level of IQ predicts success in many professions and jobs, but this is only the case up to a certain level. Once this prerequisite level is reached, it is more often emotional intelligence (EQ) that will enable someone to excel in a position of leadership or responsibility[3]. This is fast becoming a significant differentiator and a valuable asset in the world we live in today, a world where people must communicate and collaborate with larger numbers of sometimes globally dispersed colleagues about increasingly complex topics.

There is strong evidence to suggest that a child's capacity to regulate emotions and exhibit self-control, both elements of EQ, predicts long-term success in life. Here's one example: a large-scale study of more than a thousand children over a span of thirty years found that self-control, defined as the ability to control impulses, delay gratification and stay focused on goals, predicted later financial success and positive social outcomes, over and above IQ or parental social status and wealth[4]. Maybe this surprises you, particularly the financial success part, but I hope it gets you thinking that this quality of emotional intelligence isn't just a fluffy 'nice to have', it may actually improve your child's life in tangible ways. We really mustn't underestimate how vital these skills are to the success of human beings in a broader sense. Think of them as a foundation upon which both wellbeing and achievement are built.

Some of the key components of Emotional Intelligence (EQ) are **self-awareness, self-regulation** and **empathy**, also traits that have a significant impact on parenting. Expecting your child to navigate her way through modern life successfully without these qualities is akin to putting a child with a blindfold on to an obstacle course and expecting her to win the race. Bear in mind I'm not saying every human being must have high EQ, and there are plenty of examples of very brilliant individuals who have done or are doing great things without the benefits of a high EQ. However, helping our children successfully traverse their own emotional lives, along with the ability to understand those of others around them, will reap rewards in almost any setting I can think of that involves human interaction.

Self-awareness – the bedrock of emotional intelligence

Take, for example, the trait of self-awareness. Self-awareness is essentially your ability to know and observe your thoughts, emotions and behaviours as they unfold in real time. It gives you the power to be able to experience yourself from a slight distance, as though you are watching a movie with yourself as the protagonist. Self-awareness matters because it is only when you are aware of what you are feeling and doing that you can exercise any choice or control over it. Children don't develop a fine sense of self-awareness for quite some time and some don't develop enough of it at all, even into adulthood. I have met many senior leaders in organisations who cannot work effectively with their peers and those reporting in to them because they have no idea how they come across to others. People low in self-awareness can't identify or feel their own emotions in a tangible way, let alone be cognisant of the impact of their behaviour on others around them. It is not that they choose not to be self-aware or aren't trying hard enough, but that they can't do it, much in the same way as you might struggle to answer questions about the writings of Descartes if you have never studied philosophy.

Let's take a moment to think about the role of self-awareness in a child's behaviour at home. For a child to obey your instruction not to talk with his mouth full, for example, he must first be able to have some awareness that he is doing it, and then, in that very moment, exert some control to stop himself from doing it. These are complex tasks for someone who hasn't yet fully developed that part of the brain that underpins self-awareness, without which it is simply not possible to watch yourself as you do things in real time.

For a child to control a strong emotion such as disappointment or frustration, which we frequently expect them to do, requires an even greater feat of self-mastery involving not only self-awareness but also impulse control and emotional regulation. This is difficult enough for grown-ups to accomplish in a consistent way – I'm sure we're all at this point thinking of a time when we had an ill-judged outburst that we later regretted! Much of the time this is very challenging for children of a certain age, given the immature state of a child's brain. However, it is certainly possible for them to develop it to a reasonable degree over time, something that can be enhanced or diminished by the style of parenting we adopt.

Empathy can be easily compromised in children

Empathy, another key element of emotional intelligence, rests on particular brain regions and circuits that won't develop in the same way for all. Being able to take another person's perspective is not a simple matter for children at all, especially when they are very young. It is even harder for them, and for grown-ups too, to demonstrate empathy when they are in the grip of negative emotions themselves. In addition, it is difficult to demonstrate empathy when we feel threatened, unhappy or distracted, because the empathy circuits in the brain can be temporarily suppressed, for good reasons that I'll share with you later. Children often lash out at others and say hurtful things, not just because they may be low in empathy, but also because self-control, which is the psychological resource we use to restrain undesirable impulses, relies on a limited reserve of mental energy, typically powered by glucose, that can become depleted with usage[5].

In other words, if your child has had to use self-control to rein in her needs and impulses all day at school, she may have little willpower left when she comes home at the end of the day, which is one compelling explanation for why young children can be so difficult (read volatile/ deranged/making you pine for a glass of wine at 5pm) at the end of the day. This is magnified when they are hungry or tired because our moods and stresses impact our emotions and our ability to exercise self-regulation. In the same vein, if you have been busy all day at work or at home, using your mental energy to remember things, be polite to everyone, make decisions and deal with issues, you might get fatigued and could be vulnerable to the slightest provocation from your children as the day wears on. This is when good emotional-regulation skills are so important both for children and parents.

Many of us intuitively realise that our actions, words and attitudes can impact a child's personality and emotional habits in enduring ways. Parenting plays a greater role in shaping your child than providing for them materially, passing on rules about how to do well in life, and exposing them to various opportunities to do so. Strong scientific findings show us that how we connect with them, understand them and care about their inner mental and emotional lives can have an impact on the development of neural circuits in their growing brains. So, I hope to have convinced you these emotional characteristics matter but what does this mean for parenting?

Reflection:

What do you really want for your children?
Take a few moments to think about the outcomes you most want for your children. Which of these resonate most with you?

- Happiness
- To be secure in themselves (accepting themselves as they are)
- Resilience/emotional wellbeing
- Physical wellbeing
- Success – if so, how do you measure it?

- Popularity or being well liked
- Wealth
- Power
- Authenticity
- Contentment
- Excellence – winning and excelling in some domain
- To be loved and able to sustain close relationships

Now take a few moments to think about *why* these things matter to you. How much is this a reflection of your knowledge of your child's unique personality and interests, and how much of this might be a reflection of your own unmet needs, insecurities or worries? Are any of the outcomes you seek for your child contradictory? For example, success and high attainment don't always go hand in hand with happiness, satisfying personal relationships or contentment.

What are the values you want your children to adopt?

Values are the implicit personal beliefs and ideals that motivate and guide our behaviour. They are the 'under the surface' outcomes and behaviours you care about. Your values may be unique to you, your family and even your cultural heritage. Take a few minutes to think about the values you want to cultivate in your children. Here are some examples of values:

- Truth, honesty, fairness and justice
- Personal responsibility, initiative
- Hard work, tolerance of discomfort, endurance, discipline
- Integrity
- Kindness, generosity
- Tolerance and compassion for others, altruism, forgiveness
- Achievement
- Power, status
- Control, personal ambition, self-interest
- Humility
- Sincerity, authenticity, transparency

- Conformity, self-restraint, modesty
- Respect for tradition, culture and religion
- Nurturing, caring

Which of these values matter most to you? Children will learn about and imbibe these values from what they observe you doing on a day-to-day basis, from what you emphasise and praise, and from how you explain the behaviours and rules you expect them to follow. Reflect on how you demonstrate these values to your children and how they guide your decision-making. Are you displaying the very things you expect and want from your children? For example, if you repeatedly snap at your partner but expect your children always to speak kindly to people, they will sense this inconsistency and it might weaken the respect they hold for this particular value.

Similarly, the decisions you make convey your values in an implicit way that children internalise. If your child is complaining vociferously about having to do some household chores after a long day and you placate him by doing them for him, he may not come to respect the value of selflessness or even hard work and tolerance of discomfort. However, if you show little empathy when he is tired and rigidly or stridently insist that he complete his chores he may not learn the values of kindness, compassion and tolerance. If, on the other hand, you show empathy by recognising he is tired, sitting with him and caring for him whilst he complains, but compromise by gently insisting he does at least some of his agreed chores that evening, you will be showing him how to be guided by the values of kindness, tolerance of discomfort and altruism.

Criticising children when they don't demonstrate the values we wish to uphold is not as effective as positively reinforcing the times they or others do demonstrate them. Think about how often you discuss instances when people around you or in the public domain exemplify the values you admire. For example, if you value kindness, do you talk openly about kind things that people have done? The things we say, and sometimes don't say, shape how our children come to view the world around them – what they take for granted and what they appreciate, admire and cherish.

Chapter One: Key points

- As parents most of us want our children to be happy and successful in life but, by focusing on short-term outcomes and behaviours, we sometimes miss opportunities to shape the traits, values and abilities that predict resilience and wellbeing in our children in the long term.
- Good parenting is based on a strong underlying foundation of heartfelt connection. This sustains and nourishes us with a deep sense of emotional, mental and physical wellbeing.
- Many of the conditions of modern life hinder connection between parents and children.
- Emotional intelligence, and especially the ability to exercise self-control, predicts positive outcomes in later life. But children's brains develop slowly; the brain parts that are responsible for decision-making, impulse control, emotional regulation and more complex moral reasoning don't fully develop until the mid-twenties.
- Parenting with connection in mind shapes their growing brains and augments their chances of developing inner resources for wellbeing, resilience and emotional intelligence.

CHAPTER TWO

How does parenting shape your child's capacity for emotional intelligence?

The quality of the parent–child relationship, and in particular the depth of the connection in a 'felt', embodied sense, fundamentally shapes the process of neural integration in the brain. Neural integration refers to the linking of different aspects of the brain, the mind and the body. A powerful new area of research in psychology, called Interpersonal Neurobiology, shows us that when we develop coherent links between these vast and complex neural networks in the brain, we are more likely to develop the traits that underpin emotional intelligence; traits such as self-awareness and self-regulation. Interestingly, relationships play such an integral part in the shaping of these neural circuits that some experts in the area include relationships as an essential component in the definition of the human mind[6].

WHAT ABOUT GENETICS?

But hold on a minute, isn't that an awful lot of pressure for parents? What about DNA? Isn't this type of thing innate biology, genetics? Yes, your child's traits and temperament will be determined by their inherited genetic material. The temperament your child is born with will have a significant impact on how he regulates emotions, or whether he has an optimistic or a pessimistic emotional style. But genes are like light bulbs controlled by switches, and whether they are switched on or not depends on the physical and emotional environment the child inhabits. Although certain characteristics that are woven into our DNA, for example eye colour, won't change much from birth, other genes can be modified, turned on or switched off depending on what we experience: scientists call this Epigenetics. Our genes are not our immutable biological destiny.

Some characteristics are more open to change than others

The extent to which traits and characteristics can develop will be essentially limited by genes and DNA but can also be impacted upon by experience. Certain traits and characteristics have a limited genetic set range within which we can wiggle, but for others, there may be a broader scope for the influence of environment and nurturing. The heritability of a certain characteristic explains how much of the variance in a certain trait in a population can be explained by differences in genetics. Some characteristics will have higher heritability than others and in general the more complex characteristics that result from more than one set of genes will have lower heritability. As a very rough example, studies show that 40 to 60 per cent of the variance in the well-established Big Five personality traits, which are Extraversion, Agreeableness, Neuroticism, Openness and Conscientiousness, is down to genetics[7].

It is important to note that these studies are far from conclusive and just because something has lower heritability does not mean that it is open to change and vice versa. This is because heritability only tells us about how much variance in a trait across many people is down to genetics; it does not tell us anything about what causes them to be that way. Traits, behaviour and various conditions that influence emotional intelligence (e.g. autism spectrum disorders) are also shaped by brain architecture, hormones, neurotransmitters, and other factors that may or may not be open to influence. This is an important message for parents to take away because the spread of pop psychology often creates a sense that we can all be anything we set out to be, putting untold pressure on parents to hold themselves, and their children, oppressively responsible for everything the children do or don't do. This often slides towards a place where the parent–child connection becomes barren and under-nourished.

It really pays to bear in mind that our children are shaped by both nature and nurture and it is too simplistic to think that we can divide up which impacts what in a way that provides us with any certainty. However, we do know, from a very strong research base, that parents can and do have an influence on their children, especially in terms of emotional regulation, resilience, and their relationships with others, something I'll elaborate on as we go along.

EXPERIENCE SHAPES US IN TANGIBLE WAYS RIGHT DOWN TO THE CELLULAR LEVEL

So how does experience shape us and how do we, as parents, shape our children? What we do with our children doesn't just influence them through memory, learning and behavioural imitation. A child's experiences in life, however small or large, whether emotional, mental or physical, cause the release of biochemicals that travel through their brains and bodies, sending signals to their cells about how they need to change and adapt in response to that experience. This promotes experience-dependent growth in the child's brain that can persist for life. In a similar way, when the child does not experience enough of certain chemicals, for example, pleasurable, joyful, rewarding chemicals such as dopamine, serotonin and noradrenaline, the child's capacity to experience those emotions later in life may be compromised, leading to a higher risk for depression. This physically alters your child's brain and body over time and it is how experiences, including relationships, come to mould and change their very beings fundamentally at a biological level.

Your child's early emotional experiences also shape how her reserves of metabolic energy are used to support growth and restoration in various parts of the brain that are in critical stages of development in the first few years of life. If the child faces repeated stress, for example, energy reserves will be diverted to deal with those stressful experiences, taking them away from the task of building and connecting up those brain regions that might support emotional intelligence.

At the same time, your relationship with your children forms a blueprint for something called their attachment style, a characteristic pattern of connecting with others and regulating emotions. Your attachment style as a parent influences how tuned in you are to your child's emotional needs and how you help your child to manage them. It is this level of 'attunement' to your child that influences how the emotional regions of your child's brain will develop (in addition to his or her genes). I'll lay out the different attachment styles in a later chapter but for now, let's look at how our experiences shape us at a biological brain-based level.

HOW DO OUR BRAINS INFLUENCE OUR EMOTIONS AND BEHAVIOUR?

To understand how parenting influences the development of our children, it is important that we understand how 'nurture' shapes 'nature' at the level of the brain, because what we are capable of doing, or not, rests heavily on biological processes that stem from the brain and its connections with our body. At a very general and broad level, the extent to which a person will have the capacity to demonstrate certain emotional traits and characteristics depends on the brain: the size of various brain regions, the level and type of activity within them, and how well connected they are with other regions.

As I mentioned above, we are also shaped and influenced by our hormones, neurotransmitters and other chemicals, for which we have a varying number of receptors in the brain, the development of which can also be affected by early experiences. These chemicals shape what we are capable of feeling and doing in response to the situations we encounter so the number of receptors we have for a particular chemical, e.g. dopamine or oxytocin, determines how much of the chemical is available to us and how we react to it. Children who face physical or emotional deprivation, or worse, severe trauma, in their very early years, may not adequately develop the brain regions that play a part in emotional regulation. They may not be responsive enough to the neurotransmitters and hormones that promote healthy coping strategies such as seeking comfort from people and being soothed by their presence. As an example, pregnant women who have high levels of circulating cortisol (the stress hormone) give birth to babies who have a more sensitive stress response system, making them more likely to respond adversely to stress in later life. Likewise, those who are fortunate enough to have parents who are attuned and emotionally balanced in the way they manage emotions in themselves and their children, will have a better chance of developing those vital brain regions implicated in the development of emotional intelligence.

State versus Trait: Don't be too quick to label and judge

Before I give you a quick description of how brain regions and their connections shape us, I want to remind you that our behaviour can be shaped by temporary factors and more permanent ones, or in other words, state and trait. State influences our emotions and reactions in the moment and can be influenced by transient variables such as hunger, sleep deprivation, hormonal fluctuations, impulses etc. Trait-based behaviour, on the other hand, will tend to be more stable and persistent over time, attributable to personality traits and characteristics that are hardwired into us. It is important that we don't judge and label our children with negative traits when their behaviour is more likely to be a result of a transient state.

There are individual differences in brain structure, connectivity and chemistry that make us unique

From one person to the next, there are differences in the way our brains are structured. These differences make us the individual and unique people we all are. Here's a quick description of these differences at a brain level:

BRAIN REGIONS – We all have many different brain regions, which are broadly specialised for various traits, abilities and processes, so putting it in simplistic terms, they do different things in different ways. The size of these parts can vary from person to person, which results in differing levels of the trait, ability or process that particular brain part is involved in. For example, several brain regions, working together in various patterns, are responsible for decision-making, empathy, impulse control and so on. How these brain parts develop will depend on genetics and environmental influences, including early relationships and emotional experiences.

NEURAL CONNECTIONS – The number and density of connections within a brain part, and also between them, influences our emotions and behaviours. How well brain regions integrate and connect with other regions and the body as a whole can alter what we are capable of feeling, doing and thinking. Once again, there will be a genetic influence over the way these connections arise, and the time-span over which they develop, but this is also influenced by our experiences. We know that early trauma, frequent bouts of unregulated stress, and problems with emotional attunement between a baby and his or her primary caregiver can affect how well various parts of the brain grow these vital neural connections. Note that too much or too little neural activity in various brain regions may compromise the development of the traits, abilities and processes those regions enable.

If a child has not yet adequately developed a certain part of the brain that is responsible for something, it is as impossible for them to demonstrate that trait as it is for you to speak a foreign language you have never spoken before simply because I tell you to do it. Children grow into themselves in the most literal sense because the part of their brains that allows them to make rational decisions, empathise, control impulses and pay attention, develops slowly over time and is open to influence along the way. When we do not have sufficient connectivity between certain brain regions it becomes very difficult to demonstrate traits such as empathy, which requires a brain–body connection, and also self-regulation, which requires connections between certain higher and lower layers of the brain. Chastising your child for being rude in the moment may not be a compassionate thing to do because when children don't have the depth of connections between the parts of their brains responsible for empathy or impulse control, they are simply not fully capable of it. This is not to say you allow them to act with impunity, but rather that berating them makes little sense when they are not entirely in control of themselves yet.

WHAT KIND OF PARENTING POSITIVELY SHAPES THE CAPACITY FOR EMOTIONAL INTELLIGENCE?

The development of balanced emotional traits is especially dependent on the emotional responsiveness of a parent or caregiver towards the child in the first few years of life. I don't mean whether or not you feed your baby or change her nappy when she cries; yes, caring for your baby's physical needs is important, but the kind of responsiveness I'm describing here is about how sensitively you pick up on the emotional signals of your child and whether you respond to those signals in a way that matches what your child needs from you.

Because young children don't yet have the brain connections that enable them to manage their own feelings and reactions, they don't have an 'off button' for emotions and uncomfortable feelings until they are much older. Yet they are capable of feeling anxious, stressed, sad, angry, excited and more, all possibly quite overwhelming when you consider that they can't consciously calm themselves down. This means they must rely on the emotional regulation capability of their parent of caregiver to regulate and soothe them when they feel distressed or overwhelmed. This is apparent right from the start as an infant's heart rate, stress chemicals and other biological markers of nervous-system functioning come to synchronise with those of the caregiver who holds him. Even when children are older, their brains are far from complete, and they continue to need a loving, regulated and understanding adult brain to act as a 'proxy brain' for them. It is in this context that they learn to process and make sense of their experiences in a healthy way.

This is a non-verbal, instinctive process that involves what is called 'coregulation' of emotions. The parent uses his or her own emotional awareness and regulation capabilities to synchronistically regulate the emotional responses of the child, who is mostly incapable of doing so himself. This process creates a deep feeling of trust and emotional safety between parent and child, so vital for the growth of traits that promote resilience and good relationships later in life. When parents consistently fail to pick up on and respond sensitively to emotional signals from a child, some of the neural architecture that underpins emotional intelligence does not develop in an integrated way.

For this process of coregulation to happen intuitively and effectively, you need to be able to feel moved by your child rather than preoccupied by your own thoughts, feelings, needs and insecurities. When you are tuned in and connected, you feel your heart physically warmed by eye contact with your child; when you listen to your child you don't just hear the words they say, you sense what they are trying to get across and you can listen with a sense of compassion and acceptance rather than judgement or worry. It stems from an emotional system that involves feelings of stillness, contentment, connection and openness, rather than defensive, angry feelings, or feelings of preoccupation, frustration, ineptitude, shame or guilt, all emotions many parents experience in the early years. It doesn't involve thinking, talking or doing anything – rather it's a way of just experiencing the moment together, being in 'synch' so to speak. And when your brain is working in a healthy, integrated way, and your environment isn't conspiring against you, this process unfolds naturally and without any conscious thought or control whatsoever.

This process of connection, of resonance between parent and child, called 'attunement' by developmental experts, builds the foundations for emotional intelligence in a child, including the capacity for resilience and self-regulation. It does this because it leads to physical growth in brain regions responsible for these traits, but it also contributes to the process of neural integration and connectivity between these regions, enabling children to develop a healthy sense of themselves[8]. This is the bedrock of mental and emotional wellbeing and it is upon this secure base that other layers of parenting, such as discipline and authority, can be built. I don't mean to pile on the pressure but rather to emphasise that we, as parents, can shape their brains in positive ways too (we don't just have to worry about messing them up!).

It helps to see that your role is to be a proxy brain whilst their brains are work in progress, but simultaneously to shape their brains for resilience, emotional regulation and other enriching characteristics through their experiences with you. When you do this, it must be from a standpoint of openness and acceptance for who they are emerging into rather than who you want or need them to be for your own gratification or comfort.

OUR BRAINS CAN CHANGE AND DEVELOP OVER TIME

We now know that much of the foundational brain architecture that supports emotional regulation is actively sculpted in the first few years of a child's life because this is when brain growth happens apace. Your child's brain will grow rapidly in the first three years of life, not just the development of different parts of the brain that enable various traits and abilities, but the depth of connections between these regions too. This sculpting process is not all about growth either; neurons that are not used get ruthlessly pruned away. And in these cycles of growth and pruning, your child's brain slowly takes shape, moulded by your own brain and his experiences, so that he is now matched to fit with his own unique emotional environment.

Some brain parts develop slowly over time and are therefore open to influence for a longer period than others. Parts of the brain in the pre-frontal cortex area responsible for regulating emotions, impulse control, rational thought and making informed choices, are very rudimentary in childhood and continue to grow and mature until they reach the peachy age of twenty-five. Bearing this in mind, we must question how reasonable it is of us to expect our children to demonstrate the kinds of behaviours that we often require and demand of them in their early years. It may not be kind or fair of us to expect them to exhibit these traits in a consistent way, regardless of how tired, hungry or emotionally dysregulated they may be.

Please don't feel disheartened if you think you and they have missed the boat for good emotional development in the first few years. Although children experience the most rapid brain growth and development in the first few years of life, the human brain is capable of being shaped through learning and repetition over its entire lifespan. The brain is a bit like play dough; it is malleable and can change in response to new experiences, something called **'Neuroplasticity'**. Knowing this must give us a sense of optimism and perspective because we are learning that our children are by no means set in stone from early childhood. Assuming no brain injuries or other neurological/psychological disorders, we are capable of change, sometimes drastic change, all through our lives, though this will slow down in the latter stages. Also bear in mind that, although change is possible, we must accept and respect each other as we are, with all our unique characteristics, and influence our children with the knowledge that they can only change within the parameters of their genetic inheritance.

PARENTING HAS NO END GOAL

Take a pause and make a commitment to focus on the long-term picture. You are not running out of time to help your children. This is not a race to the finish line and your child is not an object that must be moulded in a specific way within a specific timeframe. You are partners in this journey for life. And to be the kind of parent who can positively shape your child's developing brain in the ways I've just described, you will probably need to change in some ways too. I know this because I've done it and it has transformed my family life and my connection with my children. But my parenting journey had to start with me, and for this reason, I hope you'll learn and change more about yourself than you will your children.

Over the next few chapters, I will show you, with fascinating and eye-opening brain science, why the ways in which we now live are obscuring our ability to feel a real sense of connection with our children, and I will show you what to do, in a very real and tangible way, to train your mind and nervous system to allow these moments of emotional resonance to emerge. Once you learn how to feel connected to those parts of your brain and body that naturally lead to good parenting, you'll find that you can rely on your intuition so much more. You'll find that your children feel so good being around you that they begin to want to please you. They'll listen because they want to, not because they are afraid of losing your love or being punished. Parenting will become truly reciprocal. Rather than feeling as though you are perpetually 'giving' of yourself with sometimes little return, you will feel nourished by your time with them because being together will feel rewarding and pleasurable. This won't be the case all the time; you'll still have moments of despair, frustration and guilt at the odd outburst, but more often than not you'll feel genuinely, 'in-the-bones' happy being around them.

It is only once you learn how to create a certain state within yourself, in your nervous system, your brain and your mind, that all the wonderful tools and techniques put forward by experts in the parenting field will start to bear fruit. You may even get to a point where you don't need behavioural tools and techniques because you learn to 'read' your child and intuitively know what is needed. This is what we humans are built for, but a number of the facets of modern life obscure our ability to feel connected, in the absence of which parenting can resemble a monumental challenge and even a source of stress. Once you learn to regulate yourself and your own emotions, parenting will seem to flow along like a calm and clear river on a beautiful sunny day, with occasional ripples from gusts of strong wind, or even a transient storm, but nothing that won't settle in a few hours or days. But first, you need to 'prime the pump' for heartfelt, connected parenting. And it starts with your brain.

Exercise to build a foundation for better attunement:

As learning how to tune in to our children's emotions requires a certain stillness, an ability to be aware of what is happening in the moment, let's take a few minutes to connect with ourselves and notice how well we pay attention to what we are experiencing. Please note all these breathing exercises are available on my Soundcloud profile and it is often easier to follow an audio guide than it is to follow a written exercise. To access them, go to soundcloud.com and type in ShellyChauhan.

Breathing Exercise: Learning to notice and feel rather than think and judge (five minutes)
This exercise is based on the concept of being mindful and learning to pay attention to the present moment. It helps cultivate awareness and acceptance of what is happening without judgement. It allows us to slow down and experience things rather than being lost in thought. Instead of 'doing' we must try sometimes just to 'feel' and let ourselves 'be'.

1. Start by finding a comfortable place to sit, where you can be quiet for a few minutes without any obvious distractions. Try to sit with your legs and arms uncrossed and your back reasonably upright. Set an alarm for five minutes.
2. Close your eyes, if that feels comfortable, and become aware of your breath moving in and out of your body. Try to notice yourself as a living, breathing, feeling person rather than a collection of thoughts and intentions.
3. You might find it easier to keep your mind focused on the breath if you count each breath from one to twenty (then start again at one) or if you silently recite the word 'in' as you breathe in and 'out' as you breathe out.
4. As you exhale, try to slow down your breath.
5. Try to bring your full attention to the process of noticing your breath. It is perfectly normal for your mind to wander and trying to stop this might be futile. The best thing to do when your mind wanders is to notice it and gently bring your attention back to the process of breathing.
6. Try to tune in and really notice the sensations within your body as you go through this exercise. Whatever you do, try not to get caught up in perfectionism because our goal is simply to relax and notice bodily sensations without judgement. There is no right or wrong way to do this.

7.   If you can, repeat this daily until you come to the next breathing exercise in this book.

Chapter Two: Key points

- How well we interact and connect with our children shapes the process of neural integration in their growing brains. When they develop coherent, balanced links between the various parts of the brain and the body, they are more likely to demonstrate the characteristics that underpin emotional intelligence such as self-awareness and self-regulation.
- Our genes and DNA shape us, but genes are not immutable; some are unchanging over the lifespan and others are switched on or off depending on our experiences.
- Children's brains can be shaped by the behaviour of their parents in different ways. Their experiences cause the release of biochemicals that alter the activity and growth patterns of cells in the brain and body in ways that can endure over long periods of time.
- Our brains are divided into multiple regions that are broadly responsible for different characteristics and abilities. The regions differ in size, as does the density of neural connections within and between them, and the receptors we have for different chemicals, all of which combines to influence our personalities, emotional habits and behaviour.
- Children rely heavily on the presence of a parent who is tuned in to their emotions to help them to regulate themselves, meaning we must act as 'proxy brains' for our children until they are capable of self-regulation.
- Being tuned in to the signals of our children requires us to be moved by them in a non-verbal, heartfelt sense.
- Although much of our neural architecture is defined in the first few years of life, our brains are capable of growth and change over time, something called neuroplasticity. Children's brains are still in active development until their twenties and can modify all through

their lives. This means parents can change and adopt new parenting behaviours, and influence and guide children towards the outcomes and values we seek for them.

Priming ourselves for connection

Connection is the foundation for the kind of parenting that develops our children's resources for wellbeing and resilience. But however wonderful this sounds in principle we know that staying connected and compassionate can be really difficult. After all, parenting can bring about an avalanche of challenging emotions that many of us may not have experienced in years, if at all. I recall how unexpectedly tough I sometimes found the parenting process when I had my first child; I just hadn't expected that I, with my supposedly calm persona, could be capable of such frustration, worry and sometimes even anger. At times, especially when it seemed to go wrong, it all just felt too personal and I lost my sense of perspective, spending hours googling parenting advice and ending up confused.

Parenting can give rise to feelings of anxiety, frustration, rage, disappointment, defensiveness, embarrassment and rejection, all of which go against the grain of what we want to experience with our children. When we experience these feelings frequently with our children, our built-in 'caring' and 'reward' circuits shut down and we struggle to derive pleasure and fulfilment from being around our kids, instead feeling resentful or stressed by our inability to control or enjoy the parenting experience. Our children perceive this, whether consciously or not, and have an emotional response to it that shapes their subsequent behaviour. For example, they might feel hurt and rejected, which, depending on how well they are able to regulate these feelings, may lead to an increase in tantrums and defiance on the one hand, or withdrawal and indifference on the other. Once this happens, a parent and child can quite easily get stuck in a cycle of annoyance and negativity towards each other.

Feeling connected promotes the capacity for experiencing empathy, contentment and joy in the company of another person, which helps mitigate the more challenging moments that every relationship entails. I am convinced, based on an understanding of how we use our brains and how this shapes our relationships, that we are more at risk of emotional disconnection than ever before.

In this chapter I will describe the three essential characteristics that underpin connection, and over the course of this book I will explain each one of them with reference to some fascinating brain science to bring it all to life. We will delve into the brain, what it does and how it has evolved in layers that each have a broadly different purpose relevant to our parenting behaviour.

I will also describe the two hemispheres of the brain, which enable us to pay a very different type of attention to the world around us; one that facilitates connection, empathy and acceptance of each other as we are, and another that renders us capable of achieving, striving, labelling and judging, all behaviours that can, when taken too far, hinder connection, and promote self-focused behaviour. Over this and the next chapter, I will lay out why I think parenting with each of these hemispheres in the lead will result in a different type of relationship with our children and how, ultimately, real connection can only stem from a balanced brain with one hemisphere slightly in the driving seat.

THE THREE ESSENTIAL ELEMENTS THAT UNDERPIN CONNECTION

These three elements have a remarkable effect on parenting – not just what you think about it and how you define your relationship with your children, but how parenting feels to you and to your children on a daily basis. They are not traits or psychological characteristics, neither are they purely behavioural; rather they are ways of being, processes or states, that arise as a result of neural connectivity between our brains, minds and bodies. They are:

- Open Presence
- Emotional Regulation
- Emotional Safety

Each one of these processes hinges on complex brain–body systems that I will help you understand and manage over the course of this book, not only to greatly enhance your relationships with your children but also to significantly lower your stress levels and increase your sense of wellbeing. But first we need to understand a bit about our brains and how and why they have evolved into their current state.

UNDERSTANDING THE BRAIN – WHAT IS ITS UNDERLYING PURPOSE?

Before we go into what the brain is, I think it is important to consider what our brains have evolved to do. At a very general and simplified level the brain receives information from our bodies and surroundings, organises that information into mental maps, interprets that information, and generates a response to it. Many of the brain's activities are designed to help us regulate our physical, emotional and mental states so we can maintain a certain survival-promoting state of inner equilibrium, for example, maintaining our core temperature and heart rate, responding to hunger and thirst signals and so on. We are generally programmed to want to maximise reward (positive sensations and emotions) and avoid threat (negative sensations and emotions) which manifests as an 'approach' or 'avoid' orientation to our experiences that plays out on a rapid and continuous loop outside our conscious awareness.

When you anticipate that something will be rewarding, you open up to that experience, either emotionally, mentally or physically, just as when you anticipate or judge something as unpleasant, you close off to that experience in defence, regardless of whether that response is rational or not. This happens in the form of rapid shifts in the state of your nervous system that lead to a cascade of other reactions, some of which you may consciously perceive (for example your heart rate going up and your muscles tightening). To interpret bodily sensations and guide your responses, the brain also needs to make connections and assign meaning to your experiences, something that might involve making predictions, assumptions and interpretations and generating stories. Again, this happens mostly outside of conscious awareness and sometimes with scant regard for truth or rationality.

WHAT IS THE BRAIN MADE UP OF AND HOW DOES IT WORK?

Your brain is a mass of soft tissue comprised of over a trillion cells, including over 100 billion neurons. Each neuron is densely connected to other neurons via roughly five thousand connections called synapses. The brain, though it is comprised of different parts that are broadly specialised to have different functions, is very much a whole and complex system in which various regions work in tandem in an inextricably interconnected fashion. The brain is also densely connected to the body via the nervous system, which is a carrier, much like a cable, for information from the body to the brain and vice versa. If you imagine a complex map with trillions of possible routes going off in every foreseeable direction, you start to get a sense of how intricate and interconnected our brains really are.

Thoughts and actions result from the firing of neurons

Every conscious mental activity such as a thought or a reaction occurs as a result of millions of neurons firing at any given moment. Neurons communicate with each other via their synapses through the passing of chemicals such as neurotransmitters, which are busy little messengers scurrying around telling the neurons whether to fire or not. When neurons fire, for example when you have a thought or read this sentence, they do so in a temporary pattern that lights up and then disperses, like fireworks exploding in the sky. However, when the same pattern of thoughts or behaviours is repeated several times the neurons involved fire repeatedly, and begin to form a more permanent trace. This makes it more likely that the same set of neurons will again fire together in that situation, so in other words, you are likely to experience the same thoughts or behaviours in that trigger situation without consciously choosing to do so. The more a certain group of neurons fire together in a certain pattern, the more likely that the resulting thought or reaction will become a habit. As Donald Hebb, the psychologist, is supposed to have succinctly put it, 'When neurons fire together, they wire together'[9].

A typical neuron may fire five to fifty times a second[10] and each neural signal that fires is a source of information to another neuron; information that exists as a form of electrochemical energy. All this information, in the form of energy, is moved around the brain and nervous system via our neurons. The process that organises and regulates this flow of energy around the brain and body is called the mind[11]. The mind is not a fixed, tangible entity that resides in any one part of the brain but rather is a dynamic process of organised energy flow across the body, the brain and its various structures. The mind, and therefore your awareness of who you are – your conscious sense of 'youness' – emerges from the processes of your brain, your body and your relationships and connections with others. The brain, mind and body are so inseparably linked that changes to one part will inevitably lead to an impact on the others.

When we think about the fact that we can self-regulate and influence the patterns of energy flow through our minds by consciously choosing to direct our attention in a certain way, i.e. choosing to have certain thoughts or do certain things, and that in turn can physically alter our brains, we begin to realise the power we hold to change our own mental and emotional habits. This process rests on neuroplasticity: the ability of the brain to physically grow and change in response to experience. Not only is it possible to grow areas of the brain that give rise to various emotional, mental and physical characteristics, but it is also possible to alter the density of connections within and between them. What I'm saying here is that you, as a parent, can consciously use your mind to gradually change your brain–body circuits in a desired way, and when you do so, you can have a transformational impact on your child, your relationships and your wellbeing.

THE BRAIN HAS EVOLVED IN LAYERS WITH DIFFERENT FUNCTIONS

This alone could be the topic of several dense tomes so please bear with me whilst I present my potted history with just enough detail to make my points about the parenting relationship. Our brains haven't evolved to this stage by chance alone but rather in a way that is uniquely adapted to the environment we have encountered over millions of years of evolution on our planet. These adaptations happen very slowly over large swathes of time, whereas industrial and technological advancements have often been so rapid as to create a deep mismatch between evolved brain-based emotional capabilities and the situations we find ourselves in today.

Through the course of human evolution our brains have developed into a three-layered structure where the layers are heavily interconnected yet retain some of their distinctive features[12]. Daniel Siegel developed a clever way to bring this to life: hold up your hand in front of your face with your palm facing you. If you put your thumb into the middle of your palm and close your fingers over it you recreate the three layers of the brain. The most primitive layer is the equivalent of the reptilian brain and is comprised of the brainstem and cerebellum, which is the visible base of your palm under your curled fingers. This is connected to the spinal cord and the rest of the body, represented by your wrist and forearm. This part of the brain regulates basic bodily functions and sensations such as breathing, heart rate, digestion, balance and some movement.

The second layer, represented by your thumb, is the mammalian brain, which evolved when mammals developed complex needs requiring more advanced circuits for their survival, in particular the ability to care for their babies. This layer of the brain incorporates the 'limbic system', which plays a pivotal part in the generation and regulation of emotions, so vital for healthy relationships and nurturing. Your coiled fingers and the upper part of your palm is the third and most recent layer to have evolved; the primate brain or cerebral cortex, and in particular the prefrontal cortex which, when well connected with the limbic system (the thumb), allows us to regulate our emotions and integrate them with logic, reasoning, anticipation of consequences and general self-control. This is the most complex layer and exerts a great influence over the rest of the brain. When these layers work well together and communicate effectively via the connections between them, we are able to parent in a loving and consistent way, without undue stress and anxiety, or on the other hand, disinterest or resentment.

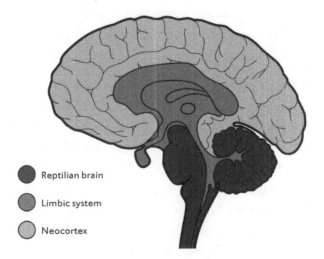

Reptilian brain

Limbic system

Neocortex

As well as being organised horizontally, your brain is also organised vertically into the left and right hemispheres, with a band of fibres called the corpus callosum keeping the two distinct by allowing one to inhibit the other (in a changeable way rather than a rigid way) and maintaining a balance and cooperation between them.

How you come to 'see' and understand your child through the lens of these two hemispheres matters greatly in parenting, especially if parenting for emotional balance and resilience. Learning about how and why these hemispheres evolved to pay attention to the world around us in different ways will help you to understand how the way you perceive and interact with your child changes the very nature of your relationship with him. The way you perceive the world via your left and right hemispheres plays a strong part in all three of the key conditions for connection – maintaining an open presence, regulating emotion and also creating a sense of emotional safety in your interactions with your child, but it has a particular impact on developing an open presence, a prerequisite for heartfelt parenting.

THE RIGHT AND THE LEFT HEMISPHERES: THE LENS THROUGH WHICH YOU SEE AND ENGAGE WITH THE WORLD

Imagine you're walking through a beautiful garden and see an exquisite and colourful flower, bathed in sunshine, that sends a small rush of happiness through you. A short while later, you see the parts of the very same flower laid out for scrutiny under a microscope, disassembled and disembodied from their surroundings. Would you feel the same about the flower? Would you even know it as the flower you marvelled at earlier? Perhaps you were unmoved by the flower in situ but are fascinated by the petals under the microscope. The way the flower is presented to you, i.e. the context in which you experience it, matters because it changes not just the way you feel about the flower but the very nature of how you come to 'know' it and therefore what you see. It is much the same with the brain and its two hemispheres. This section about the hemispheres is based on the illuminating work of Iain McGilchrist on the divided brain[13, 14, 15], through whose remarkable research I have come to believe that how we use the two hemispheres of our brains has a relationship-altering effect on the way we parent. I have summarised here my understanding of his work and the implications for parenting.

The brain has been evolving painstakingly, layer by layer, for millions of years, but it has also evolved to be structurally divided into two hemispheres; one on the right and one on the left, a division that has endured over time. Each hemisphere has within it several brain regions with differing functions, many of which are split into two parts, one on the left and the other on the right. These two hemispheres are asymmetrical; for example, the brain is wider on the left posterior area and the right frontal area; and the right hemisphere is heavier and larger than the left hemisphere. The right appears to offer greater connectivity between the different regions contained within it, whereas the left, in contrast, has greater connectivity within each region rather than between them, which gives a clue as to the key difference between these two hemispheres. We know from studies of patients with brain damage that their view of the world and even their personalities may be dramatically altered in different ways depending on whether the damage has occurred on the left or the right side of the brain.

You may have come across popular theories about what the left brain is and what the right brain is, for example, the left brain is literal, linear and logical, whilst the right brain is creative, emotional and artistic. We now know, from brain imaging studies, that the two hemispheres can't be accurately talked of as having entirely different functions from each other because both hemispheres are utilised for almost every human function including language, the processing of emotions and reasoning. But why, as Iain McGilchrist asks and answers, if both hemispheres are involved to some degree in almost everything, and if both have enough neuronal mass to sustain themselves independently of the other, do we have divided brains at all? It is not so much whether they do different things that matters, but rather *how* they do what they do because we know the two hemispheres do things in different ways. They do things in different ways because they have each evolved to pay a different kind of attention to the world. This is a profound distinction. You can only see the world as it is presented to you, and each hemisphere presents you with a different version. What you see shapes what you find. If you take in one version over the other, you cannot know what you do not see.

HOW DO DIFFERENCES IN THE HEMISPHERES CHANGE THE WAY YOU SEE THE WORLD?

If you take yourself back in time to the earlier days when our brains were evolving their basic structures, you can imagine that life was considerably simpler in some ways, whilst also being fraught with challenges such as being at the mercy of ravenous predators. We were preoccupied with two main tasks: to forage for food and to protect ourselves from danger. For the most part, we were immersed in nature, often in wide, open spaces that necessitated a heightened awareness of potential dangers. To spot an unpredictable predator we'd have needed to pay a certain type of attention to the environment around us: attention that is broad, expansive, open to any emerging possibility, and without any preconception. We couldn't have known or predicted from where a possible predator might spring so we would have had to pay attention to the whole of the space around us. This type of sustained, broad, expansive way of paying attention, without judgement or preconception, is enabled by the right hemisphere of the brain.

But on the other hand, we also had to forage for food in the undergrowth, which required the opposite of the wide, open attention we needed to protect ourselves from a predator. To notice and pick a berry from amongst the undergrowth, we had to pay a narrow, focused type of attention, much like the beam of a torch in the darkness, that would have illuminated that specific morsel of food and highlighted it as distinct from its surroundings. This type of attention is enabled by the left hemisphere. As these are very different ways of viewing the world it would have been impossible to do them simultaneously with one unified brain. However, with a divided brain it became possible to pay both types of attention concurrently or to switch rapidly between them.

Interestingly, the left brain controls the movement of the right side of the body, including the right hand, which we most often use to grasp and manipulate objects. The left hemisphere also controls most of the function related to verbal language, making it the part of the brain that enables us to 'grasp' things – both verbally and physically. To get an idea of how the

39

left operates, imagine you have spotted a berry in the undergrowth and need to identify what it is and whether it is safe to eat or not. You would do this by analysing its individual features and matching them with what you already have stored in your memory system. It's blue, it's round, it's small, ergo it is a blueberry. In keeping with this, the left hemisphere, in its expanded role in the world we now inhabit, reduces down, analyses, and fits what it sees with what it already knows or wants to see.

When we generalise this way of attending to the world, we can see that the left is narrow, precise, focused and controlling in the way it takes in and interprets its environment. It is the opposite of open because it seeks to reduce things down so it can obtain them, make sense of them, categorise them, and in using them as it desires, ultimately control them.

Parenting with connection requires the ability to switch between open attention and focused, narrow attention. Imagine one of your children has spoken rudely to another and you want to correct this. If you become lost in a narrow, detailed focus on the perceived transgression of a social rule, you might snap at the child who perpetrated it with little awareness in that moment of whether your child feels hurt by your tone of voice and whether your reaction is in proportion to the behaviour of that child. You will lose sight of how your child is feeling. On the other hand, if you are able to switch your attention flexibly, you might correct your child, something that naturally involves judging and labelling (left hemisphere), but you might also intuitively watch her face to see how she is feeling about what you are saying and how you are saying it (right hemisphere). You might automatically soften your tone if you notice that she is feeling ashamed or sad, and in doing so, you will respond in a kinder, more proportionate way.

WHAT ARE THE MAIN DIFFERENCES BETWEEN THE HEMISPHERES AND WHAT DO THEY MEAN?

The left hemisphere is about striving, grasping, narrowing and controlling

The left hemisphere is primarily linked to grasping, striving, mastery, achievement and pleasure, and is more closely associated with the neurotransmitter dopamine, a chemical that is released in the brain in anticipation of a rewarding experience. It underpins our desire to approach things, to seek novelty and stimulation and to some extent, to experience pleasure and excitement.

Unsurprisingly, because the left seeks to narrow our attention in order to grasp things, both physically and mentally, it strives for control and certainty and will favour certainty over complexity, subtlety and nuance. The left hemisphere is primarily concerned with analysing a representation of something so it can match its features with what is familiar and in doing so, make sense of it. It is not capable of taking in something in all its living, breathing, feeling 'realness'. It enables us to take a step back from the world, to strategise about it, to 'know' it in a technical sense, but not to actually experience it. If you are analysing something, you are not actually experiencing it.

The left brain is not directly connected to the sensations of the body, and given real experiences, including love and connection, are mostly sensory in nature, all experience is presented to it second-hand, having already been processed by the right brain, a fact that it is, rather frighteningly, unaware of. As it is made for narrowing down its focus, it has no appreciation for real 'lived' experience, or any sense of the 'wholeness' of something including the context in which it is occurring; rather it tends to see things as isolated incidents complete in themselves. Just as an animated film builds the whole from tiny static moments, so the left brain assembles its sense of the whole from details that it weaves together until it has something it can recognise.

This means the left hemisphere is not as effective in the processing of emotions, which tend to be complex, ambiguous, multifaceted, sensory experiences that arise in the body. The left is primarily concerned with grouping and categorising things, for which purpose it has to ignore uniqueness and focus on generalisable features that allow it to give something a label and make sense of it. This is how stereotyping and other forms of categorisation arise and may obscure our empathy for people as individuals. When we view the world and people through the lens of the left, they become like objects to be analysed, labelled, controlled or used, all so we can 'grasp' the mettle of them and get the sense of certainty and satisfaction that labelling brings with it. Viewing people or things through the lens of the left hemisphere, including our children, alters how we perceive them.

The left hemisphere creates a sense of yourself as a distinct, autonomous being, with a set of rules and expectations for how others 'should' treat you and what kind of life you 'should' have. It is associated with your conscious, focused will relating to yourself – what you believe you deserve or want; what you need; how you want your life to unfold; your aspirations; your need for comfort or autonomy; your goals and all those things that when unmet or violated create emotional discomfort. This is spectacularly unhelpful in the context of parenting because being a parent means you are now inextricably intertwined with the emotions and experiences of another human being who is entirely dependent on you and will be for many years. When you allow yourself to accept and simply experience this, and all the natural emotional and physical discomfort this entails, without judging it or trying to control it, parenting becomes easier almost straight away.

The right hemisphere is the location of your implicit, social, emotional brain

The right hemisphere, because it is connected with the body and receives all the signals from the gut, heart and other organs, is connected with our sense of intuition, instinct and emotion. It allows us to feel and process social and emotional signals from our surroundings, including facial expressions, which is the basis for connection both with humans and with the world around us.

The right brain is designed to be open to whatever is emerging just as it is, without a rush towards analysis or assumption. Because it is able to appreciate uniqueness and novelty, it is about taking in the whole experience of something as it arises without any judgement of what

is being presented or any need to explain what it sees with reference
to what it already knows. As an example of when this ability may be
compromised, I have come across many parents and corporate clients
who, looking at things through the lens of the left, cannot understand
and accept that someone could have an emotional reaction that is wholly
dissimilar to one they would have in the same set of circumstances.

The right hemisphere is not associated with physical or mental striving,
but rather connecting with things – with their characteristics; the
sensations they evoke in us; and their relationships to other things
around them. This is very different to the left, which can only know things
in the context of what they 'should' be. My son, with his left-dominant
hat on, often made me laugh with his comments about the weather; if
he had read the weather forecast for the day and his observation of the
weather didn't quite match his expectation, he would say, in a vexed
tone, 'But it should be sunny!' He struggled to grasp that the weather is a
natural phenomenon that does not comply with man-made rules for his
convenience and comfort.

The frontal part of the right brain is responsible for empathy, intuitive
moral behaviour, impulse control and social connection. The left
hemisphere cannot decipher and process emotional information from
people's faces as well as the right. Almost all emotional arousal and
felt emotional processing happens via the right hemisphere, though
conscious analysis, mental processing and control of emotion can be
left-hemisphere dominant. Most negative emotions such as fear, shame,
sadness – which, interestingly is linked to the capacity for empathy
– and vulnerability, lateralise to the right side of the brain, except for
anger, which evidence suggests lateralises more to the left. There is also
evidence that the happiness and pleasure associated with achieving goals
stems mostly from the activation of the left side [15] and that other mild
positive emotions may be more strongly associated with left activation
rather than right but there is some controversy and disagreement in
this area so we cannot know for certain. It is plausible that conscious,
'me-related' emotions may lateralise to the left, including some that may

play a part in the resilience and happiness that stems from unrealistic optimism, grandiose notions about oneself or even self-confidence.

How we view problems and challenges varies from left to right

How we view problems and resolve them is also dependent on how we use the two hemispheres. Whilst the left hemisphere is involved in calculations, the right can help us perceive the meaning of the conclusions the left may have arrived at. The left values rules and concepts, for example, thinking about anything in the abstract; in terms of what it 'should' or 'could' do, whereas the right hemisphere values real, felt experience. The left is uncomfortable with ambiguity; it invents plausible narratives to ensure that it can retrofit the information it is presented with into some predictable pattern of existing knowledge. It can manipulate the facts and create illusions to back-fill the gaps in its understanding of things, without being able to see that it is doing so, making it a master of self-deception and delusion. Though the left may be clever it is not necessarily truthful, especially about itself and its own responsibility for mistakes.

When viewing problems and generating solutions, the right hemisphere is able to present several possibilities and can remain open to various options until an acceptable solution is discovered, unlike the left, which fixates on a viewpoint that appears to validate its pre-existing assumptions about things. It has the ability simply to deny or discard information that doesn't fit with how it wants to view things, something called Confirmation Bias. The left is prone to unrealistic optimism and, perhaps because of this, cannot adequately handle severe stress. Whereas the right hemisphere may be realistic about itself and what it doesn't know, the left is somewhat delusional and has a positive bias in the way it evaluates itself. Unlike the right, it is not able to cope as well with change, ambiguity, flexibility, emotion and intuition, all the very things that children, with their uniqueness and unpredictability, bring into our lives and need from us.

When you look at the world through the lens of the left, you see things as objects to be understood explicitly, including your children, whose feelings and thoughts you cannot perceive implicitly, right brain to right brain. The left gives you control over your world (and your child) and the right connects you with it, two things that may come into conflict in the parenting relationship. We, and our children, may now be more skilled at talking about and analysing our emotions but I question whether we are losing our ability to notice and experience the real, physical sensation of emotions.

Here is a summary of some of the differences between the left and right hemispheres:

Left hemisphere characteristics	Right hemisphere characteristics
– The left sees the parts – Narrow attention and scrutiny – Judges what it sees according to how things 'should' be – Expects information/things to fit into a system or pattern that makes sense – with preconceived ideas of things	– The right sees the whole – Broad, open, sustained attention – Takes in what it sees with openness to what is actually happening – Experiences things without judgement – Can appreciate uniqueness without needing to fit what it sees into a system, pattern or category
– About grasping, expecting, controlling, striving, achievement – Prone to unrealistic optimism and self-delusion	– About 'being', connecting, acceptance of what is emerging, openness to things as they are – Capable of vision and novel thought
– Favours certainty, rules, categorising and labelling – May be prone to 'black or white' thinking – Prone to confirmation bias because of the tendency to fit what it sees with what it wants or expects to see	– It sees the interconnections within and between things/people/our natural world. – Can tolerate ambiguity and uncertainty – Can hold and accept multiple explanations – Manages complexity, including paradox and contradiction

– Not connected to the sensations of the body and has no 'felt' sense of things – only a verbal, analytical sense – Can't decipher facial expressions, emotions or social signals well – Not especially empathetic, though can be capable of cognitive empathy (knowing how others do, or might, feel – different to being able to feel moved by it)	– Connected to the sensations of the body – Dominant in the processing of emotional information, facial expressions, social interactions, empathy – Favours real, lived experience over abstract analytical information-processing
– Only understands things explicitly, in a verbal sense – It has no sense of what it doesn't understand	– Understands things at an intuitive, non-verbal, implicit level – Can have a sense of what it doesn't fully understand or know and is capable of wisdom

OUR RELATIONSHIPS BECOME DISTORTED WITHOUT THE WISDOM OF THE BALANCED RIGHT HEMISPHERE

Think for a moment about how this applies to our relationships with our bodies these days; our bodies no longer seem an integral part of 'who I am' but rather a collection of parts to be judged, controlled and altered to fit with the rules we've internalised about how bodies should look and perform. The prolific rise of body-image disorders, selfie-taking, cosmetic surgery, fanatical fitness trends and diets, performance enhancement amongst young men and a whole host of other issues in the younger generation is testament to what happens when we engage with ourselves through the lens of the left hemisphere.

Body-image issues are on the rise in young people and it is a tragedy that so many of them aren't able to feel a sense of acceptance and connection with their bodies but rather concern themselves more with what their bodies can and should do for them. The body has become objectified to the point where it is acceptable to tinker and tamper with it continually, much like one would do a machine, to conform to a set of perfectionistic standards that often deny the reality of the human body

– that it is diverse, ever-changing, vastly complex, lumpy, hairy, smelly, unpredictable and full of so-called imperfections. None of these things matter when we feel a strong connection to and respect for a person as a whole, embodied, emotional being. It is only when, via the lens of the left, we objectify the person or the body, that we are able to be repulsed or alarmed by small features in isolation that are quite insignificant when taken as part of the whole.

Self-criticism, which we know plays a part in depression and anxiety in children and adults, also stems from this left-brain tendency to view oneself against a never-ending list of 'shoulds' and 'musts' rather than simply accepting oneself as a unique, complex, feeling, breathing human being. Children seem increasingly driven by what they think they should do rather than how they actually feel about things. We, as parents, need to understand how the rules we set up for our children might gradually lead them to ignore their inner needs and feelings.

The level of discombobulated scrutiny that the left is able to generate is unhealthy in the context of human relationships, whether with ourselves, our partners or our children. When you take a whole, break it down into its parts and analyse it, you lose your connection and intuitive sense of that very thing. Love is a good example of this; the more you analyse it and rationalise it the less you feel it in a visceral heart-warming way. You can't read a book about emotions and become more emotional as a result, especially if you are left-dominant. Even if you learn to feign the emotion it has little benefit because humans are most responsive to felt, visceral emotions that arise in the body and are processed implicitly and non-verbally via the right hemisphere.

If I'm making it all sound like the right is superior and the left is inferior, that isn't quite what I'm after, because we know that we need both hemispheres working well together to function effectively. Not only that but the right and left both may be compromised in their growth and fail to function effectively. For example, when the right hemisphere has not developed soundly, it may lead to deficits in emotional regulation, empathy and even morality. Parents with an overactive right hemisphere may struggle to find parenting pleasurable and may be overly defensive, protective or rejecting of their children. Depressive feelings and feelings of threat and anxiety stem from imbalances that lateralise to the right hemisphere. We'll come to the role of the right hemisphere in emotional regulation in later chapters but for now the point I want to make is that we need both hemispheres to function well together for us to be balanced, compassionate and wise parents.

Should the right hemisphere be in the lead?

Coming back to discussing the hemispheres at a broader level, as Iain McGilchrist argues[17], we would ideally lead with the right so we had a sense of vision, wisdom, connection and perspective, with the left as a strong back-up to operationalise, analyse and provide the mechanisms through which we go about striving to make things happen. But for thousands of years, and more recently with the advent of consumerism, technology and scientific advancements, events suggest we have become more left-dominant in the way we attend to the world. The education system, and particularly our focus on verbal language, explicit learning, rules, detail, analysis and machines that have allowed us to master and control our experiences, all serve to heighten this left-dominant way of being in the world. The ability to grasp and manipulate the world to suit our purposes has reaped huge benefits and led to feats of advancement for humankind that would not otherwise be possible, but given our increasing dependence on this way of being, is not without its consequences.

The more we focus on achievement, comfort, control, certainty and pleasure, the less connected, empathic and wise we become, not least because we are so busy doing things and striving for things that there is no time to truly notice anything, especially our internal sensations – the language of emotion and connection. Worryingly, because it controls so much of our verbal capabilities and therefore conscious thought processes, the left hemisphere is able to achieve dominance over the right, with the right receding to a point where it may take a back seat, ending up silent and unnoticeable.

Some of us may be more left-dominant in our parenting styles whereas others may be too led by the right. Some may find themselves flipping between the two; between rigidity and laxity, being too lenient and empathetic on the one hand or too dominant and judgemental on the other. What we are missing more and more these days is the balance between the two. As a result, there are profound and powerful consequences for our physical, mental and environmental health. This imbalance between left and right dominance is changing the way we parent, and in turn, may be changing the way our children's brains develop, which over time will change the very nature of humanity. We only need to think for a moment about the role that machines and technology play in our lives these days to see the endless possibilities for this shift in how we relate to each other. In the next chapter, I'll outline how I think the hemisphere imbalance is affecting our ability to parent well and why that matters so much to our children.

One of the more worrying aspects of all this in relation to parenting is that when we start to relate to our children through a left-dominant mode, we start to judge them as deficient in ways that deny the reality of who they are. We become angry and frustrated when we feel they are not doing things in line with what we think they 'should' be doing. Parenting books inadvertently foster this because once you have read or imagined a certain outcome from either yourself as a parent, or from your child, you will feel as though something is missing or is deficient when things don't turn out as you'd envisaged. For example, if on reading this book, you start to think that your family life should be characterised by calm, connected parenting, that your children 'should' start to listen to you, respect you and demonstrate emotionally regulated behaviour, however well intentioned you may be, you will have slipped into parenting primarily from the left rather than the right hemisphere.

There is no reason why your children should be anything other than what they are in that moment. If they knew a better way to be, and their hormones, chemicals, nervous system and brain were able to support this superior state, they would have adopted that more optimal behaviour. The point I'm making is that connected parenting will flow when you try to accept your children and yourself and your family life as they are at that point in time without demanding they be different simply because you know it is possible. This fixation with controlling our experiences and judging them against a stream of 'ideals' in our minds stems from the left-hemisphere way of attending to the world. Parenting with openness requires you to give up the demand that your children be a certain way and cultivate the ability to be emotionally regulated and compassionate with them, regardless of whether they are meeting your or others' standards or not. Once you do this in a heartfelt, authentic way, and I must be clear that this is not a thinking or talking state, but a feeling state, they will start to calm down, listen and respect you.

Exercise: Connecting with your right and left hemispheres

Start by finding a comfortable place to sit, where you can be quiet for a few minutes without any distractions. Try to sit with your legs and arms uncrossed and your back reasonably upright. Set an alarm for seven minutes. You might find this exercise challenging at first, and if so, please don't give up or become frustrated. There really is no right or wrong way to do this and each time you do it, you will build the pathways to do it more successfully in the future – remember that neuroplasticity is based on repetition!

1.   Close your eyes and become aware of your breath moving in and out of your body. Over the next five or six breaths, try to notice the sensation of breathing without any sense of judgement or striving.
2.   Now as you inhale, inhale up into the space above your eyes, on either side of your forehead. When I say inhale up into this space, I mean take your full attention and focus there as you inhale. Try not to get lost in visualising the breath but focus instead on how it feels.
3.   As you exhale, try to exhale into the same space but focus on

softening and slowing down each exhale so it is longer than your inhale. Notice whether you find it easier to focus on the left or the right side. Once you are comfortable with taking your attention to the space behind the eyebrows and forehead, you can move on to the next part of the exercise.

4.   Continue to inhale into the same space but on the exhale, try to take your full focus to your right hemisphere. If you can focus on the whole of the right side that is fine, or if you can focus only on the frontal area that is fine too. Remember to keep your exhalations very soft, gentle, relaxed and slow. Progressively allow your face and body to soften and let go of tension. Do this for at least two minutes.

5.   Now repeat step four on the left side.

6.   Now inhale and exhale into the left and right side together, trying to connect with, and pay attention to, your whole brain during the exhalation.

7.   Try to do either this exercise or the one at the end of Chapter Two daily until you come across a new exercise. If one feels particularly calming and restorative, practise that one more often. Let yourself be guided by what feels most soothing, though bear in mind that sometimes what you most need is not necessarily what feels the most comfortable.

Reflection: Use these questions as a guide to think about how the way you pay attention to the world around you, via the left or right hemispheres, alters and affects your parenting style

Being and experiencing (right hemisphere) versus achieving and controlling (left hemisphere):

•   How much of your time with your children is focused around activities you think you should do with them (taking them out, playing games, feeding them, reading to them) as opposed to connecting with them in a calm, still way (sitting together, relaxing, listening, cuddling, eye contact)?

•   How often do you allow yourself to just 'be' with your children without getting caught up in judging or correcting their behaviour?

- How often do you feel really moved by your children, intuitively sensing their inner feelings and needs?
- Are you able to connect and experience shared emotional experiences with them non-verbally, through eye contact and facial expressions?
- How often do you allow your children to run around and play outside freely without much concern for cleanliness and timeframes?
- How frequently do you do spontaneous things as a family, or allow them to engage in unplanned play time with friends, neighbours, etc?
- How much time do you spend connecting with things that are not related to achievement and consuming, for example, art, music and nature? Please note I don't mean doing these things because you think you have to, or are afraid of missing out, but really being moved by them.
- When your children do things that upset you or others, are you able to take a step back and see things from their perspective before telling them off or coming to a conclusion about the causes of their behaviour? (balanced right and left)
- How much of your parenting relationship is about getting them to abide by rules, for example, food rules, etiquette, timings, sleep rules, social rules, and how much flexibility are you able to show when they don't want to, or aren't able to, follow those rules?
- Do you sometimes find yourself fixated with controlling the outcomes your child may face in life and becoming anxious when you cannot predict how things might turn out?
- Do you often miss opportunities to notice their emotions and feelings, tending to focus on whether what they are saying or doing fits with your notions of right and wrong?
- Do you often find yourself struggling to understand them, seeing them as unreasonable, difficult to fathom or too hard to handle?

Chapter Three: Key points

- There are three essential conditions for real connection to emerge in the relationship between a parent and child – these are open presence, emotional regulation and emotional safety.

- We are at risk of disconnection because of the ways in which our brains pay attention to, and engage with, the world around us. To understand how to remedy this we need to learn about our brains work and how to generate a sense of balance and integration.

- Our brains primarily seek to interpret our experiences and organise our responses to them, in order to keep us safe and maintain a state of equilibrium within our bodies. We are generally hardwired to want experiences that feel positive and to avoid or defend ourselves against experiences that feel threatening or negative.

- Our brains are made up of billions of neurons, millions of which interact each time we do or think something. Neurons can wire together, forming habitual patterns of thought or behaviour that are likely to arise automatically when we face situations similar to those we have already experienced. But we are also able to use our minds to consciously direct our attention towards certain thoughts and behaviours that can bring about desired changes in ourselves. The ability of the brain to change in response to experiences is called neuroplasticity.

- Our brains are organised into three key vertical layers but also into two hemispheres, one on the left and one on the right. Both hemispheres are involved in most of the mental activities we engage it, but they enable us to do those things in different ways because they are each designed to pay a different kind of attention to the world around us. This has a profound impact on the parenting relationship and, in particular, parenting with an open presence.

- The left hemisphere enables us to pay a narrow, precise attention to things and fits what it sees with how it believes things 'should' be. It is about grasping, controlling, judging, analysing, categorising, labelling and ultimately manipulating what it encounters in order to make sense of it and use it. Parenting with a left dominance can lead to rigidity and a loss of perspective.

- The right hemisphere pays attention in a way that is open and accepting of what is occurring, without preconceived ideas or judgement. It is connected with the sensations of the body and is dominant in the processing of emotions, social signals and 'felt' empathy. It senses the interconnectedness of things and can, when functioning healthily, promote a sense of vision, wisdom and perspective. Heartfelt connection emerges primarily from a right-brain state of open presence.

- Some of us may be primarily left- or right-hemisphere driven in the way we parent, and others may flip between the two. Good parenting requires integration and balance within and between the left and right hemispheres, with the right hemisphere ideally slightly in the driving seat.

- Many of us are now more left-dominant in the way we function. This has an impact on mental health, wellbeing, and our sense of meaning and purpose.

CHAPTER FOUR

How the two brain hemispheres shape your ability to connect

Interesting as this information about the brain and the hemispheres might be, we're here because we want to understand more about how we parent, so how exactly does the way we view the world affect our parenting habits?

If we put it in a nutshell, the left hemisphere's way of paying attention is primarily about 'grasping' something, i.e. talking, analysing, judging, achieving, controlling and planning, whilst the right is broadly about 'being' i.e. sensing, feeling, connecting, intuition. The left generates expectations (and anger/frustration when they are not met) whereas the right allows us to observe and accept things, and people, as they actually are in that moment. As authentic heartfelt connection only emerges from a (regulated) right-brain-led state of open presence, we need to parent with the right in the lead to help our children grow those important neural connections that underpin emotional regulation. More importantly, we need to parent from the right to build a truly loving and compassionate foundation for the parenting relationship, something our children need for their wellbeing, resilience and sense of contentment.

DOES CONNECTION OCCUR NATURALLY?

You might be thinking, surely if connection were so fundamental to our mental and emotional lives, it would occur naturally? Well, when we think about the fact that for most of our human evolution, we didn't have verbal language but instead relied on the language of emotion and intuition that arises in the body, heartfelt connection was probably more of a given in those days than it is now. But along with the evolution of this wonderful higher human capacity for abstract thought, verbal dialogue, planning ahead and generally dealing with things that aren't grounded in real sensations or current events (i.e. the incessant mental to-do list) we are now getting so lost in the world of what 'could be, might be and should be' that we are less and less able just to be present without judging, analysing, controlling and worrying. This explains the growing popularity of meditative methods such as mindfulness that teach us to do what brains have known how to do for centuries but has been lost in this day and age of being perpetually busy: simply to be present in the moment without a sense of judgement or striving. We need to relearn how to let ourselves and our children just be, not all the time, but for at least some of the time, because connection only emerges from openness, from tiny moments of stillness.

Table manners: An example of left- and right-led parenting in action

Let's take table manners as a light-hearted example of where our left-hemisphere rigidity might lead us astray and how we might seek a better balance. A five-year-old child, with an understandably immature sense of self-awareness, sits at the table feeling excited about the plate of food presented to her. She is ravenous and eager to eat and as she has little self-control, something that is not of her making, she immediately grabs the food with her hands and starts to chew it enthusiastically. You, in your preoccupied and somewhat critical state of mind, snap at her to close her mouth and take her feet off the chair. She feels a little surprised and affronted but closes her mouth.

A few minutes later, she is chomping loudly with her mouth open and some morsels of masticated food have migrated to her cheeks, chin and clothes. You become irritated and use a snappy tone, perhaps even use the word 'disgusting'. Your mouth, whether you notice it or not, may have curled upwards on the left side (the expression of disgust) and your tone of voice will have an edge, however subtle, of sharpness to it. She feels ashamed and because you are judging her, in that moment she no longer feels connected with you. After all, she is simply trying to taste her food and enjoy it. She is living in the moment, obeying her body's signal to devour food, and is not thinking of abstract unfathomable rules about what she can and can't do whilst she eats.

When you react in this way to the child, your child has become, in your mind, an object to be controlled; in that moment you have little connection with her feelings about her food or herself. You might not notice how her face falls when you tell her off; perhaps you won't notice anything unless her reaction to your comments becomes loud and histrionic. Your sense of how your child feels is secondary to your need to get your child to comply because she is failing against an ideal. This is a simple example of how parenting with a left-hemisphere dominance can unnecessarily create distress or rupture connection. Our children can tolerate a fair degree of all this when there is a strong and abiding sense of warmth, understanding and closeness between you. However, if there is more correction, control and misunderstanding rather than acceptance and warmth, the balance will shift and your child's behaviour will go one of two ways: defensive and defiant, or passive/indifferent, neither of which will help develop those vital emotional intelligence skills we know we now need.

If, on the other hand, you were parenting with an integrated, regulated brain, your reaction to your child in exactly the same situation might look quite different. You might not feel disgusted or irritated at all, light-heartedly reminding your child to eat with her mouth closed. Or you might still want to snap but you would quickly notice your initial reaction of irritation and disgust, stop talking before you unleash it upon your child, relax your body (it will have tensed up outside of your awareness) and take a mental step back, all this happening naturally and quickly. You might look into your child's eyes and feel a sense of the real person there; just a little, hungry child being in the moment. You might have an inner sense that children are not born with an innate understanding of why these rules matter, nor the memory to consistently recall all these rules in situ. With a kind, self-aware stance, you might help build this capacity in your child over time by teaching her why these rules matter. Adopting a mindset of curiosity rather than judgement, you might ask her **why** we insist on people closing their mouths when they eat.

After all, why have we evolved all these rules about what we should and shouldn't do at the table? Let's not forget that for most of our evolution we were out in the wild, clubbing each other hard on the head for food. And now, in really very little time at all in the scheme of things, we insist on these rigid rules that almost deny ourselves the right to have bodily sensations or emotions and to sometimes act on them. I believe most of our social rules such as eating with our mouths closed may be primarily about showing consideration for other people because it isn't very pleasant to see half-chewed food in someone's mouth.

Tenets of social etiquette are often constructed to enable us to avoid experiencing things that might disgust, hurt or upset ourselves and others. But over time we've lost the message about consideration and internalised the rule in an undeviating way that often overrides common sense. Table manners are now a symbol of something; their absence a sign of bad manners, itself a sign of something broader about social status. So, perhaps these rules came about out of consideration for others, but we've now forgotten their purpose and impose them so rigidly as to treat our own children with little consideration in the process. In some cultures, food is something to be enjoyed a little more expressively and freely without any pressure or scrutiny. I'm not advocating abandoning etiquette entirely, just relaxing a little so that our children can engage more fully with their senses of touch, smell and taste.

If you want your child to treat others with consideration, empathy and understanding, it is more important that you help nurture those brain pathways by showing kindness to your child, rather than by expecting the child to internalise endless rules about how to treat other people that they may later go on to apply in quite a formulaic way. Of course, once you've shown such consideration to your child, you do need to emphasise the rule and will probably have to continue to do so until the child internalises it (this takes patience!). Remember, it's all about balance.

PARENTING WITH CURIOSITY RATHER THAN JUDGEMENT

When we parent with the left brain in the lead, we expect certain things of our children without accepting that they are complex human beings with their own traits, needs, fallibilities and ideas that may or may not match ours. We view them as objects in the eyes of ourselves and others, to be judged, influenced and even controlled, to avoid the possibility that anyone else might judge them or us negatively, or that things might not go our way and we may experience emotional discomfort. We expect them to do things like feeding and sleeping in certain ways because we have read books that indicate children are capable of this, and when they don't conform with our expectations, we try to find ways to coax those behaviours out of them. It's hard to remember that just because we discover something is possible doesn't mean it is an imperative and should or will happen, though this is exactly the kind of thinking that left-dominant attendance may promote.

Imagine you were born with a permanently attached yellow lens in front of your eyes: you would see the world as tinged with a yellow hue and would only ever be able to see it in this way. You would not know that the world exists in a spectrum of colours. It is much the same with viewing our children through a certain frame of reference; once we label, judge, scrutinise and comment on what they are or are not, we come to expect that particular behaviour from them and what's more, because of confirmation bias, we are more likely to explain their behaviour in light of our existing assumptions and judgements. We struggle to accept deviation and individuality when it doesn't fit with how we want or need to see our children. Sometimes it feels more comfortable to label and know what we can expect from them, than to accept that there might be chaos and uncertainty in store along our parenting journey.

Social comparison ruptures connection

All this parenting information we are now presented with heightens social comparison, something we are hardwired to engage in anyway. We compare our children with other children and become alarmed or frustrated when they don't demonstrate similar traits or behaviours. We worry about them missing out on things because other children are doing them and what's worse, because of the rise of social media and technology designed to 'connect' us, we now have the capability to know what every child everywhere might be doing. When we parent with the left hemisphere in the lead, we are 'doing parenting' and it can lead to a constant cycle of negativity and stress in the parenting relationship. We are missing the point entirely, which is that what children most need to grow well is right-led openness, compassion, warmth and acceptance backed up by left-led rules, expectations and solutions, all held together in a fine balance. It is this way of being nurtured that builds those parts of the brain that lead to the positive outcomes we so desperately seek for our children.

Being misunderstood leads to anger and disconnection

One source of intense frustration, anger or sadness for children is when we misattribute their behaviour to causes that don't make sense to them, or fail to capture their intentions accurately. This happens when we interpret their actions in light of some pre-established explanation we may have, either of children in general, or this child in particular, and we view them through that lens regardless of other equally valid explanations for their behaviour, which we often do not even bother to seek. Most of us will realise we do this frequently and that we often do it in a negative direction, for example, a child who is just looking for comfort because he can't regulate his own internal state is labelled 'clingy' or a child who lies is labelled 'manipulative' when in reality he may have learnt that telling the truth leads to an explosive reaction from his parent that he finds frightening and wants to avoid at all costs. This tendency is fostered by the increasingly left-dominant world we inhabit, but it is also exacerbated when we are mindlessly busy, stressed or otherwise emotionally dysregulated.

When children feel misunderstood in this way, they become angry and resentful, because not only is the label often unfair and inaccurate, but you're supposed to be on their side and it ruptures the connection between you when you judge them. Typically, these judgements involve a generalisation of one instance of something into a more pervasive habit or personality trait, e.g. 'You're so unhelpful! Why can't you just help out when I ask you to, like your sister does?' Of course, the child may have some personality traits that make her less inclined to help unless there is some personal benefit, but it could also be that she is just six years old and cannot control her impulse to do something more pleasurable. Or it could be that she was preoccupied with something, or just wasn't feeling like it at that particular moment. Not feeling like doing something may not be a reasonable enough explanation for you to avoid something as an adult, but for a child with an immature brain, it looks pretty compelling! I don't advocate giving in to this, just that we need to be more understanding and less judgemental. It's more constructive to find ways to teach your child that helping out makes our home life work better for us all.

This type of rupture stemming from mislabelling, criticism and negative evaluation is often a cause of so-called 'bad' behaviour. I've noticed a few times over the past year that when my child seems to be pushing against me and parenting starts to feel like hard work, I have usually slipped into a negative and judgemental mindset, subtle but persistent over a few incidents, often without knowing it until I see it mirrored back to me in my child's behaviour. This tends to creep up on me when I'm busy, sleep-deprived, or have some sort of work or social event looming up that requires mental or emotional energy. When I slip out of a mindful, compassionate state into a preoccupied, disparaging state of mind, my children's behaviours, which frankly haven't changed at all, start to seem more annoying and negative than usual. Where the week before I may have thought their noise levels were an example of children being playful and joyful, in a more stressed or mindless state I might find myself thinking they are disorderly or annoying. They haven't changed – I have, or at least my inner state has. And when they sense this, and hear my criticisms, they feel disconnected from me. If this persists over even a few days, they start to become less eager to help and listen.

Being judged alters self-image and self-acceptance

Turning an incident like a lapse in table manners or a momentary loss of self-control into something that can be generalised shapes your child's sense of who he is and sets the scene for how he behaves in the future because it becomes self-fulfilling.

When a child is repeatedly criticised, he comes to internalise a view of himself as a 'defective' child. This won't be a verbal memory but more likely a shameful inner feeling he retains about himself. Once he believes he has this in his nature, he will no longer feel compelled to please you because it will seem almost inevitable to him that you will disapprove of his behaviour. He may become argumentative and defiant because he frequently feels misunderstood and comes to expect this from you. He might start to jump to negative conclusions about your comments, even when they may not be intended to be critical. On the other hand, it may even create a habit of striving for perfection or needing approval in order to avoid negative evaluation. Parenting needs to be a space where judgement is cast aside because labelling children is at best unhelpful and at worst damaging to the relationship. Our children need to be respected as complex, ever-changing, fallible and unique human beings in their own right rather than as little objects to be watched, criticised, applauded and controlled.

Even positive labelling of traits and characteristics, such as 'lovely', 'cute', 'pretty', 'sporty', is a type of objectification and generalisation. Wanting to grab a child because he is 'so cute' is not the same as feeling a loving connection. Of course, it is natural to do this some of the time because we want ways of talking and thinking about them and need the handy shortcut that labelling provides, but we must be mindful of how we use these labels because they frequently don't reflect reality. The reality is that every human being is an enormously complex amalgam of traits, values, attitudes, skills, habits and so on, many of which are changing and developing all through our lives. We are simply too amorphous to be judged and labelled and made static as we are when we are labelled. It is acceptable to judge a behaviour but not so helpful or even accurate to classify and label a whole person. It is healthier to speak about your child's behaviour rather than the whole person and in doing so, you relieve them of the pressure to keep up and allow them the option to be themselves without fear or shame. But please bear in mind that even frequent criticism or analysis of behaviour can result in a rupture to connection, and children often generalise from their behaviour to their whole selves anyway.

PARENTING REQUIRES FLEXIBILITY AND OPEN-MINDEDNESS

Our children are emotional beings because they are primarily right-brain dominant in the first few years of life, which is why there is so much emotion bursting forth at every opportunity. It is our left brains that create the labels and interpretations for why they do the things they do because we'd rather have some explanation for their behaviour than live with the ambiguity of not having any. We've become so accustomed to having control over our worlds that we struggle with the notion of randomness and spontaneity, which defines much of the early behaviour our children may demonstrate. We invent a narrative (often only loosely connected with evidence of a good quality) about our children based on our own limited knowledge because we can't know what we don't know. This type of perfectionism in our expectations of our children can go awry because children are going to be anything but predictable, and however busy and challenging our lives may be, we need to release them from the burden to provide us with comfort and certainty just because we cannot handle the pressure or anxiety of not knowing how things might turn out for them in later life.

When we are in extreme left-brain mode, it is evidenced by a fixation with control, perfection, predictability, judgement and comfort, without a sense of flexibility and appreciation for randomness, human fallibility and ambiguity. Some parenting books inadvertently encourage a left-brain way of interacting with our children and even when they are full of good advice, I have found from personal experience that I inevitably became so focused on 'doing' something with my child or looking for signs that it was working that I compromised the parenting connection in the process. Examples of this may be getting fixated with baby routines and prescriptive behavioural techniques for handling children, all of which are generalisations and may ignore the uniqueness of each child, his or her feelings in that situation, and the very context around the behaviour that is challenging. Most of these techniques are based on psychological theories that are meaningless when taken out of context, which is that there are thousands (I'm not exaggerating) of theories each with good explanatory power that you won't be aware of, or thinking about, when you are reading about and implementing the one that happens to be in your mind at that point in time. Rather than thinking 'What should I do?', we need to focus first on the question 'What do I feel?', and only then get to the business of controlling behaviour. This is something we need to encourage our children to do more of because it connects them with their sense of who they are.

I can admit I was so desperate (and rigid) in my desire to get my first baby to sleep through the night so I could return to work feeling even half as competent as I had felt before childbirth, that I lost my ability to tune in to my child's feelings from 6pm onwards. I'd read prescriptive books about routines and so on which made it seem to natural to expect a baby to go to sleep because the clock said it was a certain time. It certainly affected my connection with my son because I would go through the motions of his bedtime routine quickly and efficiently to avoid the dreaded 'tired but wired' time of the day, and although I succeeded in getting him to sleep really well, as he's grown older he has become somewhat obsessive about timings generally and struggles to relax when there is a deadline or a time we have to be somewhere. These days, the tables have turned and he is always rushing me so he can be early for things. Although we're mostly on time, he thinks I'm too relaxed and wishes I'd demonstrate a more committed approach to time management. As someone wise once said, you reap what you sow!

HOW TO ADOPT A REFLECTIVE, BALANCED APPROACH TO DISCIPLINE

If you have a genuinely warm connection with your child, you'll find it easier to accept that your child is fallible, that she will get many things wrong, and will go through difficult phases throughout her life. When you are focusing on the long term, you don't need sticker charts, naughty steps or other prescriptive methods of controlling her behaviour, though you may find some of those helpful as a prop from time to time. My advice is to use these methods judiciously with younger children if you feel you must, but not if they shame your child, are based on isolating your child when he or she is distressed, or if they fracture the connection between you.

Rather than becoming frustrated by trying to enforce endless rules whilst your children search relentlessly for ways to circumvent them, help them to understand three fundamental aspects to the rules you are emphasising.

1. Why do these rules exist and how do people feel when they are broken?

This is all about identifying the underlying reasons behind the evolution of the particular rule. It is about thinking about how other people will feel when we don't follow the rule and what impact this will have on our collective wellbeing. It is often quicker and easier to neglect the 'why' and focus instead on frightening our children or making veiled threats about what might happen to them if they don't follow instructions. Very young children do tend to follow rules out of fear and may lack the empathy and reasoning skills to engage in this type of discussion with you, but as the children get older do help them to understand why these rules exist and therefore why there are consequences for breaking them.

For example, if getting to school on time is a source of tension between you and your child, rather than shouting, coaxing, controlling or annoying them, ask them *why* being on time is important. What happens to other people when a child is late to school? How do the teachers and other children feel when someone comes in late? Try to explain that it is not considerate to keep people waiting or interrupt the activities or concentration levels of a group of people who have started an activity on time. Help your child to think about what would happen if every child came to school when he or she felt ready to leave home rather than at the allotted time.

How would your child feel if other children sauntered into the classroom at will? Life would be chaotic and we would not be able to rely on each other or organise shared experiences such as school! Remind your child that many of the rules we impose on them are really about demonstrating consideration for others so that we can coexist in a mutually beneficial way.

2. What are the consequences for you as a family and how things feel between you?

Explain how it feels for you to have to keep coaxing your child to do something that he or she doesn't want to do and how that creates a pocket of stress and tension between you. Ask her how she feels when you are having to keep telling, asking and demanding that she do things. Tell your child that it feels better for all of you when she respects what you are asking her to do, especially when it is for her own benefit. Note this only works when your demands are reasonable, not overly controlling and you are generally flexible about your children's behaviour (for example, your child doesn't feel that it is always about you getting what you need or want).

3. What are the consequences for your child of not following the rule?

Going back to the example of being late for school, ask your child what will happen when she arrives at school late and how she will feel. Explain that this is a choice your child is making and if she is going to make the choice to be late, she needs to be prepared to accept and tolerate the consequences of doing that.

A note on behaviour management, discipline and consequences

If your child cannot understand these things because he or she is very young, please be patient and persistent. You may need to temporarily cajole and control to get your child out of the house whilst she matures and grows. You may even need to use some positive reinforcement technique such as a sticker chart. However, once you child is old enough to understand the consequences of not listening to you, you might like to practice tough love. Be compassionate but do not shy away from allowing your child to face the consequences of defying the rules. For example, if your child is refusing to listen and leave home on time, and you have tried positive reinforcement techniques to no avail, explain to your child that you only have two options: one is to coerce your child into listening to you, which you do not want to do because it doesn't feel loving or respectful, and the second option is to let your child learn for herself what happens when she arrives late at school. Tell her it is her responsibility and as she is making a choice to leave late by not listening, she must be ready to tolerate the outcome of that choice. Try not to use a threatening tone of voice – this is not a way of punishing your child. You must remain compassionate but firm. You'll have to be ready to see it through and tolerate the discomfort of knowing that your child (and you) may be judged and even punished in some way.

Sometimes, when they continue to repeat the behaviour that you are trying to correct using the connected compassionate approach described above, you might need to agree some form of consequence with them for the next time it happens, for example, depriving them of something they enjoy or might be considered a privilege. I am not an advocate of using this approach rigidly but I do see it can be useful with younger children at times and there is good evidence to show that due to our inbuilt negativity bias, punishment (i.e. losing something) leads to faster learning than rewards (gaining something)[18]. Sometimes you might need to take things away to help children internalise a new code of behaviour. Begin from a starting point of trust but if you have agreed something together and they fail to abide by it more than twice, it is reasonable for you to explain that your trust weakens each time this happens, and without adequate trust you cannot give them the freedom to do certain things or have certain privileges. Discuss what they need to do to remedy the situation and give them your full support in achieving it.

If you are going down this route, try to have the consequence match the situation. So, if your child has not treated another child well, you might tell your child that you need to be able to trust that he or she can control her behaviour around others and because of this, she won't be able to have the playdate that has been planned for later in the week. Or you might take away something they value and enjoy because they must learn that in life we generally have to work and earn our privileges, and there are consequences to making poor choices. It is best to ensure the punishment is reasonable and seems fair to your child and, where possible, to make it clear in advance so they know what the consequences are likely to be.

A note on finding a balance when imposing a punishment or deciding consequences

The punishment must never feel like the withdrawal of love, approval and kindness – neither should it invoke intense anxiety or stress. You don't have to frighten them to earn their respect and you certainly don't need to appease them. If a child is crying excessively it is a sign that you have gone too far and you might need to soften your approach. On the other hand, if your child is defiant, belligerent, and can't make eye contact with you, these might also be signs that you have taken it too far, in tone or content or both. You can remain kind yet firm and authoritative in your tone and it is important that you consistently follow through with agreed consequences. To do this successfully, you have to be able to tolerate the emotions your child will experience when you impose the punishment. I often show my kids that I know how it feels when they are deprived of things they want and how sorry I am that it has come to this. I explain that I don't like doing this but it is about trust. For me to trust them to listen, they need to demonstrate that they can.

I understand you may have followed advice like this before and felt disappointed when your child repeated the forbidden behaviour the following day, week or month. This is not to say these methods don't work but may instead be a reflection of your child's immature brain. Unless your child is neuroatypical or has an underlying condition that makes attention, impulse control, empathy or self-regulation difficult, when you give it enough time and hold your nerve, it will work and it will be worth your patience and effort. These disciplinary tactics are much more likely to work without secondary side effects on your child's sense of himself and his relationship with you when there is a firm foundation of connection and warmth in place.

DEVELOPING A RIGHT-BRAIN-LED OPEN PRESENCE

As the wonderful work of the pioneering scientist, Allan Schore[19, 20, 21], reveals, the right hemisphere is capable of implicit, non-conscious understanding and processing of emotion, without verbal analysis or dialogue. This implicit emotional system develops in early childhood. Schore[22] states, 'If a baby is adequately cherished, soothed, stimulated, and respected by receiving attuned response during the first 2 years of life, the right brain – the relational, emotional, social, somatically grounded side – becomes a healthy regulator for the more individualistic motives of the left brain.' It is the process of being loved in a way that is spontaneous, reciprocal and synchronistic that puts in place the mechanisms for other vital human capabilities and positive states such as creativity, curiosity, joyfulness and excitement, affecting the very nature of how rewarding your child finds life.

Schore explains, based on the work of Donald Winnicott[23], that we can think of maternal love as demonstrable in two forms, quiet love and excited love. Both of these types of love sculpt the child's neural circuits for the regulation of various states of emotional activation and excitability – from the ability to be quietly focused or still to the ability to be highly emotionally charged. Quiet love happens when the mother and baby are wordlessly tuned in to each other and is characterised by a slow, deep intimacy. It is a state of complete openness and even vulnerability in the moment. Breastfeeding, cuddling, rocking a baby, stroking, gazing at each other, smiling into each other's eyes – these are all forms of quiet love, when baby and mother feel like one entity and the baby's and mother's brains and bodies physiologically come into synch. Excited love is more of an active state, when mother and child share attention towards something, when there is an energy and playfulness to the quality of the interaction. The mother and child take an active interest in each other and derive mutual pleasure from their engagement. Both these forms of love make the child feel validated – I am loved, I am seen, I am a source of joy and pleasure, I am whole, I am alive.

Feeling love and connection in this way is a right-brain-enabled embodied experience and not a left-brain cerebral one. Eye contact and speed of speech are two examples of how this works in action. When we're interacting primarily from the left, we tend to look at the other person's mouth or elsewhere whilst they or we are speaking, and our speech is geared towards a goal, be it correcting someone, expressing a view or opinion, changing someone's mind, or simply getting the words out so we can get back to our thoughts about other things. There may be eye contact but it is superficial, as though looking at the eyes rather than into them. When communicating in a calm, regulated state with the right hemisphere in the lead, we look into the eyes of the other person and can sense what they might need or feel, a process that generates more of a 'we' feeling than an 'I' feeling. Note this will happen at an intuitive, sensory level rather than a verbal level; in other words, you don't have to be able to articulate what you sense to have a feeling of connection. This plays out in our facial expressions, bodily movements and tones of voice – not just in our words.

Through the lens of the right, we become open to really listening and noticing what the other person is trying to get across, and more importantly how they feel about what they're getting across, even if it doesn't fit with our plans, expectations and needs. When we speak, the pace may be slower, our tone of voice might soften and demonstrate greater tonal variation, and often there will be a palpable feeling of being in synch with the other person. There is no inner sense of striving towards or against anything. In such moments you slow down; your child is not an obstacle to you, or a source of frustration for not complying with your expectations in some way, or even an interruption to the endless monologue or 'to-do list' playing out in your mind. When you look at your child, you may notice the light in her eyes, the expressions on her face, and when you notice these things with your full attention, you may have the capacity to feel really moved by what you see. Being moved like this generates a physical feeling of warmth, or compassion, around your heart. Because you have no expectations and are open to whatever might emerge, you feel relaxed, calm and accepting, a state that encourages a reciprocal sense of calm in your child. This is the basis of a strong and influential parenting presence. It makes your children want to listen to you.

Real connection happens when we slow down

Good parenting requires integration between the left and right hemispheres to create a balanced state of open presence. We are aiming for high warmth combined with high authority but with warmth and empathy in the lead. Take a moment to think about how often you really look at your child and share a conversation without having other things on your mind, without rushing, analysing, correcting, or perhaps listening only to 'cherry pick' what you want to hear? To connect with your child from your right hemisphere to her right hemisphere you need to create the space and time to see each other and tune in without any feeling of pressure or need to control anything. This emerges most strongly when the mind is quiet and still.

You'll know when you're in this state because you'll have a sense of having nothing else to do or think about in that moment; your heart rate and breathing will be variable and relaxed, you'll be looking into your child's eyes when you speak, and you'll be speaking more gently than when you're busy and have other things on your mind. It'll feel comfortable and easy and you'll find you might naturally mirror back what you're hearing in a way that is heartfelt rather than the application of a rote-learned technique for repeating back what your child is saying to demonstrate empathy. A true, felt sense of connection doesn't emerge from saying or doing the right things, and it doesn't flow when we're stressed, busy or out of synch with our own bodies.

I wonder if we're all just too busy these days to generate the pauses and stillness required for this right-brain-based connection to emerge? The difficult thing to comprehend is that it only surfaces when you stop striving for it. Sometimes, though probably not often enough, I sit down on the sofa and deliberately let go of everything I need to do and all that is on my mind (note, this takes a bit of practice at first!). I simply sit there and watch the children, softening my face and letting the tension melt away. In those moments I let myself forget about the dirty dishes around the sink, the food that has made it to the floor during the spectacle that is dinner time, which child needs what for school the next day, or those unanswered client emails lingering in my in-box. I just sit there, still and quiet, and I let myself enjoy just 'being'. What is wonderful about this, and so naturally reinforcing, is that my children are drawn to me like magnets in this state; they sidle up to me to cuddle and relax. They start telling me things that have happened to them during the day. I feel as though they soak up this feeling of openness and calm because they need it. Interestingly, they don't come to me when I'm sitting down on the sofa appearing to be present but my body is tense and my mind is busy – they sense when the moment is right.

For some of us who may not have experienced attuned and compassionate parenting in our own early years, it may be difficult to relax enough around our children to connect in this way. My primary aim with this book is to help parents switch from a 'defence and stress' mode, and even an 'achievement and pleasure' mode, to 'calm and connected', something that hinges on the right-hemisphere-led brain–body connection that I will bring to life in the next few chapters. I am going to tell you less about what to do in this book and more about *how to be* because when you get to a place where you can notice and calm the internal state of your mind and body, parenting will flow more intuitively and easily. This is the foundation upon which parenting tools and techniques designed for behaviour management can be implemented.

Over the next few chapters I will show you how our brains affect the way we regulate our emotions and as a result, how our children come to feel about being around us. This alters their willingness to be honest and respectful of our rules, and it influences how they listen to us. But through it all, we must prioritise connection, not because it leads to other outcomes we desire, but because it is valuable in and of itself.

Exercise: Connecting with your right hemisphere and slowing down (eight minutes)

To tune in to your brain and how effectively it works for you as a parent, take some time to still your mind and connect with your right hemisphere. This may be harder for some than for others but with persistence and regular practice you'll start to notice and feel the benefits. Set a timer for eight minutes.

1. Sit in a quiet place and take a deep inhale so you feel your ribs and upper back expand. Exhale and let go of tension in your upper body. Repeat this a few times.

2. Now keep your inhale and exhale soft and relaxed. Try to lengthen your exhale so it is longer than your inhale.

3. Imagine you are breathing kindness and gentleness into your brain and body with each exhale.

4. As you inhale, take your full attention and focus to the space above your eyebrows on both sides of your forehead.

5. As you exhale, focus your attention on the region of your brain above your right eye, as far back as you can, all the way to the right-hand side of your brain above your spinal cord, almost as though you are sending your breath to that area. Allow your exhale to slow down and try to connect with your body and brain as you do.

6. Once you have done this for a few minutes, try to pause for a count of two or three between the inhale and exhale.

7. As you exhale, soften your face and especially all the tiny muscles around your eyes and mouth. Let your face become relaxed.

8. Now exhale into the whole brain and head, including your face, and try to slow down and feel a sense of stillness. If you are feeling downhearted or confused about some aspect of your child's behaviour, try keeping an image of your child's face in your mind and ask yourself the question, 'What feels like a reasonable and balanced way of approaching this?' You may not find an answer but try to listen to your instincts and tap into the wisdom we all have the capability to demonstrate with practice.

9. Notice how you feel during this exercise. Are you able to still your mind and body and slow down? Do you find yourself feeling irritated or trying to resist it? Or do you find yourself feeling relaxed and still?

10. Try to practise this exercise on a daily basis. Any other exercise designed to help you mindfully and non-judgementally connect with your brain and body will also be helpful.

Reflection:

- For the next few days, take the time to notice how fast you speak to your children, especially when you are telling them off, have things on your mind or are generally busy.
- Notice whether you make eye contact with your child, especially when you are in the grip of a negative emotion like annoyance or frustration. Do you look your child in the eye and notice her feelings when you are telling her off?
- When do you become most unreasonable and critical with your children?
- Notice how often you judge, label or criticise your child, whether directly or indirectly.
- Notice what happens when you show acceptance and curiosity rather than judgement.

Chapter Four: Key points

- The left hemisphere promotes a focus on 'controlling' and judging whereas the right hemisphere enables a state of 'being' with a focus on feeling, connecting and accepting. As heartfelt connection emerges primarily from a state of being still and open to what is happening without judgement or striving, it is important that we learn to balance the way we pay attention to the world around us.
- Once we come to internalise a certain view of our children and how they 'should' behave, we become judgemental, engage in social comparison, and are critical of them when they don't conform with these expectations. This promotes confirmation bias (seeing what we expect to see).

- We often generalise a few instances of our children's behaviour into a personality trait or a prediction about what might happen in the future, which can shape how they come to view themselves. When children feel judged and misunderstood, they become angry and resentful and feel disconnected from us. This makes them less inclined to listen to us and to care about pleasing us, which is when parenting can start to feel like an uphill struggle.

- Most of the social rules we impose on our children are based on an underlying philosophy of showing consideration for people around us so that we can coexist comfortably, but in rigidly compelling our children to meet them, we lose consideration for them as fallible human beings whose brains are still in development. We need to help them understand the underlying reasons why these rules exist and therefore why there are consequences for breaking them. Help your child to understand the consequences in terms of how people feel. Enable your children to understand that they are making choices to do or not do certain things and those choices come with consequences they must be willing to tolerate.

- There are times when we might use positive reinforcement tactics such as sticker charts, especially with young children, but don't become reliant on these.

- If you must impose a punishment, don't isolate your child (by sending her away), or enforce it in a way that excessively shames your child. If you decide to deprive your child of something as a means of punishment, try to match it to the situation and make sure your child understands that you are not trying to hurt her feelings or to get back at her somehow for breaking a rule. Explain that you must feel you can trust her to conform to the rules and boundaries you have agreed before you allow her to have certain privileges.

- What children need from their parents, above all, is right-led openness, warmth, acceptance and empathy, backed up by left-led authority, rules, expectations and practical solutions. Adopting a reasonable and balanced approach to discipline will emerge more naturally and intuitively when you are connected.

CHAPTER FIVE

Parenting for connection involves understanding emotions

In the previous chapters, I described how there are three key states that promote connection between people, and in particular the type of deep synchronistic connection that allows parents and children to bond. These three states are open presence, emotional regulation and emotional safety. I briefly ran through how the brain has evolved in vertical layers and how it is also divided horizontally into two hemispheres, each designed to pay a different kind of attention to the world, with a growing and concerning imbalance in the way we use them.

Connected parenting stems from integration between the two brain hemispheres, with a strong leaning towards the open presence, empathy and acceptance generated by the right, with the left being employed to furnish us with the necessary rules, systems and boundaries that children and adults need to guide their behaviour. In this chapter we'll begin to understand the second condition for connection, emotional regulation. We'll explore what emotions are and what emotional regulation is at a basic level, delving into it further in the next chapter.

As emotions are contagious between people, your emotional state has a significant impact on your child's emotional state, as does his on you. More importantly, whilst you may have the ability to calm your emotions and restore your inner sense of equilibrium, children develop this very slowly and must rely on you to help regulate their emotions in the interim. How well you're able to do this shapes the development of self-regulation circuits in their growing brains, setting the scene for how they deal with emotions and relationships throughout their own lives. I hope, by the end of these next few chapters, you'll be as convinced as I am of the powerful impact emotional regulation has on your parenting style and why it matters so much to the wellbeing of our children.

YOUR PARENTING STYLE IS SHAPED BY YOUR BRAIN

Parenting rests on certain parts of the brain that have an impact on how we feel about and react to our children, reactions that reciprocally shape their minds and behaviour. As Hughes and Baylin put it[24], 'Children's brains thrive when interacting with adults who have the capacity to love them unconditionally, experience joy from being with them, pay close attention to them, and understand them deeply. In short, nothing is more important to parent–child relationships and to children's development than the health of parents' brains.' If you're interested in nurturing your child's mental and emotional wellbeing, you first need to start with yourself, in particular, your 'in-the-moment' ability to notice and regulate your emotions, thoughts and actions.

Emotional regulation is a vital part of parenting for many reasons that I'll outline in this chapter but not least because as parents we need to regulate our children's emotional reactions with them until they are capable of doing so themselves. This is called 'coregulation'. The way in which we coregulate them sculpts their brains, minds and relationships for the future, as our brains are designed to mould and adapt to what we are repeatedly exposed to. This parental coregulation is harder than it seems because effective coregulation doesn't come naturally to many of us. All too often, we can find ourselves influenced by our own early experiences that have shaped us for defensive, controlling, angry or anxious behaviours rather than the balanced, calm and warm approach we'd like to demonstrate with our children.

Because our children unconsciously mirror our inner emotional responses to things, sometimes the easiest way to transform your child's emotional behaviour in a particular situation is to change or better manage how you feel yourself at the time. But I can't emphasise enough that this doesn't happen at the verbal, behavioural level that many parenting approaches address; you may read, do and say all the right things to them but it is the internal state of your body and nervous system that your children will register and respond to, something called emotional contagion. Emotional contagion is an effortless, intuitive process that occurs when emotions in a group, or between two people, become synchronised through a process of neural resonance. When your child observes a certain expression of emotion on your face, similar neurons will fire in your child's brain, which may lead to her feeling a similar emotion. This is one of the proposed underlying mechanisms of empathy. But we also come to coordinate our inner states, including heart rate and breathing, through other mechanisms that I will describe later in this book.

This right-brain-led intuitive and implicit process is powerful and occurs outside of conscious awareness. Children are especially sensitive to non-verbal signals of emotion and their stress levels can be affected by seemingly small shifts in the tone of voice of an adult, or changes in facial expressions that we may not even be aware of ourselves. Your children will sense and respond to your inner emotional state and will often mirror it back to you in their behaviour. To better coregulate our children's reactions, we, as parents, need to learn how to notice, interpret and work with the sensations in our bodies and nervous systems because this is the real language we communicate with, and although it is unspoken, it is powerful. As emotions are such a vital force in organising a person's sense of themselves, their relationships and reactions to things, this is where we'll start.

WHAT ARE EMOTIONS?

Most of us intuitively know what an emotion is but defining it can be surprisingly difficult and there are experts in the field who wouldn't agree on how best to do this. *An emotion is a shift in the physiological state of a person in response to input from its internal or external environment. This shift results in a set of biological changes in the body and brain that may feel either positive or negative in tone, and may impel us towards action.*

Emotions are primarily shifts in body and brain states

Emotions begin as a response to something that your brain detects; the trigger can be something in your surrounding environment or it can be something generated from within you such as a thought. Many of our emotional reactions occur in response to things we don't consciously register. Emotions typically involve alterations in your heart rate, breathing, muscle tension, facial expressions, eye gaze, tone of voice, hormones, neurotransmitters, attention, and general nervous-system functioning, the initial emergence of which you cannot usually control, but some of which you may notice. For example, when you feel anger, your heart rate increases, your breathing becomes shallow, your muscles tense up, the corners of your inner eyebrows lower towards your nose and your lips purse together. There will also be changes in your brain via the activation of certain regions and the release of specific neurotransmitters that enable you to move into 'fight' mode.

Whether or not you notice them, these changes in your internal state, however small, may generate a sense of wanting to do something, even if that something is just to fully notice or close off to what is happening around you. It is possible to exert conscious control over these shifts in your inner state after they have started to unfold but this depends on whether you have developed appropriate connections between certain regions in the brain that play a part in the regulation of emotion, something the next chapter will highlight. Fortunately for us all, though these circuits develop slowly over time in childhood, there are ways to develop the capacity for better emotional regulation even into adulthood.

The emotional experience involves three key elements

An emotional experience will involve bodily sensations, feelings and cognition. Bodily sensations are the physiological changes that occur in your body and brain involving your heart rate, breathing, shifts in muscle tension and various chemicals that are secreted that can lead you to experience, for example, tingling and butterflies in your stomach. Once emotions begin to unfold in a series of physiological changes, they are then perceived, mapped and interpreted in the brain and mind, something the neuroscientist Antonio Damasio calls feelings. Feelings then, are the private, conscious, mental experience of an emotion[25]. Examples of feelings are fear, anxiety, excitement, shame, irritation, terror, awe, joy, disappointment, disgust and so on.

Emotions are usually accompanied by some form of secondary cognitive processing in the form of images, thoughts, interpretations, labels and evaluations of situations, which may occur rapidly and out of our field of awareness. These images and thoughts about what our emotions mean to us, about us, or about the world, arise out of mental habits such as beliefs and attitudes that are shaped by memories of how the experience of that same emotion has unfolded in the past. Do bear in mind that our thoughts and interpretations are not necessarily grounded in facts or evidence but are uniquely personal to us. In the case of your children, they will experience bodily sensations but may not know what they are feeling until they are older. When they are under two years of age, they are not likely to have any thoughts accompanying the emotion but they will begin to as they age.

For example, if someone steps in front of you in a queue when you are short on time, the three elements to your emotional experience might unfold like this:

Bodily sensations – tension around your shoulders and lower face, pursing of your lips, slightly elevated heart rate, slightly accelerated breathing.

Feelings – irritation and concern.

Cognition – 'This is unfair. Some people just don't know how to behave! I'm never going to get there on time.'

The thought about people behaving badly links with the feeling of irritation and the thought about not getting somewhere on time is associated with the feeling of concern.

Here is an example of how a child may experience the emotion of shame. A child who is criticised or handled in an abrupt or brusque manner may experience shame, which results in intensely uncomfortable sensations in the body, particularly around the heart and gut, and involves a sudden drop in heart rate. Depending on his age he may or may not know he is feeling ashamed.

These sensations make the child react in one of several ways: he may want to avert his gaze and shut down so he cannot really feel what is happening to him (freeze), or possibly to lash out or speak up in self-defence (fight) or to cry and run away (flight). If the child is not helped in managing these difficult sensations and feelings, the child may assign a negative meaning to that experience; perhaps in a series of thoughts, inaudible to himself, around the theme of, 'I'm a bad child', 'I'm not good enough' or 'I'm unlovable'. This whole episode, as a fusion of images, bodily sensations, feelings and thoughts, becomes embedded in the child's memory system so that a future episode of criticism, or even a hint of forthcoming criticism, may evoke a similar response. These memories exist as a 'sense of things' rather than as specific memories that can be narrated easily. Over time these memories build up into a lingering sense of oneself, but not usually in a way that we are able to easily access and articulate, at least not without a finely honed habit of self-reflection.

Emotions alter how we perceive things

Emotions shape and organise our experiences, i.e. they alter how we react and respond, and because of this, they affect perception, memory, attention and various other cognitive functions[26]. In other words, our emotional reactions change how we pay attention, how we think about things and also therefore how we react to them. Your embodied emotional state will prime you to respond to your children in different ways. For example, when you are in a stressed or agitated state, something as simple as a child asking you questions, however innocent, may be perceived as an infraction of your mental space, and you might find yourself responding with irritation. However, if you are feeling relaxed and content, you might view your child's questions as a sign of curiosity and reward her for her interest with a descriptive and engaging answer.

There are different types of emotions with physiological signature patterns and triggers

There are various categories of emotions, ranging from positive to negative in tone, for example, sadness, anger, disgust, fear, surprise and happiness, some of which are considered to be universal. Each of these emotions has a distinct signature of facial and physiological characteristics[27], underpinned by specific neural circuits. Various emotions not only have unique blends of physiological patterns but they also differ from each other in terms of the preceding events, or trigger situations, that typically evoke them. Each emotion has a broad theme that tells us something about the trigger situations likely to lead to it; the closer a situation is to the central theme, the more likely it is that the situation will arouse the emotion that corresponds with that theme.

Anger, for example, is primarily about being thwarted from achieving a goal. The theme of anger is, as Paul Ekman, the famed psychologist dedicated to the study of emotions, puts it, 'Someone interfering with what we are intent on doing'. Anger can also be ego-defensive, i.e. we become enraged when we perceive we are being attacked physically or psychologically. Many people can react quite defensively to criticism or perceived rejection, sometimes even from their own children. Children themselves can become angry and defiant when they perceive that our criticisms or treatment of them is unfair and violates their expectation of being cared for and supported. Fundamentally, anger may be a defence reaction against not only interference with what we are aiming for, but also a defence against feeling physically or emotionally vulnerable. Children frequently experience intense frustration or anger (those dreaded tantrums!) when they believe they are being blocked from doing or getting what they want. This is not a sign that your child is 'bad', 'naughty' or 'out of control', but it may signify that your child struggles to tolerate the emotion of disappointment or the frustration of not getting her way. It is a sign that she needs your help in learning how to calm herself in such situations.

Emotions may occur together

Whilst emotions have different triggers and patterns, they don't always occur in the singular; various emotions may emerge and run together, which adds to the complexity of trying to stay calm and sane in the midst of all those intense sensations. Anger, for example, may arise with fear and even sadness. A 2011 study of the different sounds children make during tantrums revealed that **anger** (screaming and yelling) and **sadness** (crying, whining, fussing) surfaced and faded in continuous waves during the tantrum so that their respective audio signatures seemed to occur simultaneously[28]. This is quite revealing, as often tantrums are thought of as being a manifestation of rage, though in reality they are interlaced with sadness, something parents may not pick up on and respond to. There is often a disconnect between what we feel on the inside, what we reveal on the outside and the meaning that others around us attribute to our emotional expressions. Each one of us, as parents, brings our own internal stories to this process, often interpreting children's emotions in a way that fits with our own emotional state, beliefs and expectations about the world. Flawed as this process is, we are not made to see through it easily and it is this 'jumping to conclusions' about our children's feelings that most often leads to a rupture of connection between us.

Emotions are different to moods

Emotions are different to moods. Emotions tend to be transient but moods often persist over longer periods of time. For example, you may feel the emotion of irritation once but if you experience it repeatedly over a period of time it may be that you are in an irritable mood. Emotions usually last between a few seconds and a few minutes but certain emotions, like sadness, may come in waves over a longer period of time. Mood can be influenced by hunger, sleep duration and quality, nutrition, negative life events, and a whole host of other factors. Being in a certain mood can be a trigger for certain emotions.

As parents, I'm sure we all know the impact a mood can have on our emotional reactions. I, for one, know that regular sleep deprivation, of even just a few hours, puts me in an impatient, restless mood that compromises my levels of tolerance with my children and makes me more likely to assign a negative explanation to their behaviour when an equally valid and more compassionate explanation may exist. Exercise and meditation, on the other hand, can calm and stabilise my mood, making me feel open, warm and accepting of my children, even when they are bickering or being what I might otherwise label 'difficult'. How I feel at the time changes how I explain their behaviour, which changes how I react to it.

THE PROCESSING OF EMOTION OCCURS MAINLY IN THE RIGHT HEMISPHERE

We've so far established that emotions arise as a series of physical shifts in our brains and bodies, which we then perceive and map in the brain, giving rise to feelings. Bearing in mind that the right hemisphere alone is connected with the sensations of the body, it performs a primary role in the processing of emotions. As Hill succinctly puts it[29], 'The right brain is dominant for the primary, automatic, unconscious processing of emotions. The left brain performs their secondary, conscious processing in words. The right brain processes affective (emotional) information first.' This is very significant to parenting because many of our reactions are driven by implicit processing of events between us and our children, something that is deeply shaped by our own childhood experiences, housed in the right hemisphere.

For some parents, their reactions may stem from too dominant a left-brain orientation to parenting. The left hemisphere, as you may remember from Chapter Three, tends to put a certain spin on things, mostly in a self-serving way designed to fit the information it receives with what it already knows or wants to see. If, for example, you have a belief that children/people can be manipulative and devious, you are likely to see your child's difficult behaviour as a sign that they are trying to get the better of you, rather than the result of an immature brain struggling to deal with things in the best way it knows how. This means that rather than observing and feeling empathy for your child based on an open-minded acceptance of what is occurring, you will be reacting to your own 'take' on why they are behaving in a certain way, and may feel angry or frustrated. On the other hand, if you have a strong right-hemisphere bias, without adequate integration between the hemispheres, you may become overwhelmed with empathy for your child, or may experience too much anxiety and worry when your child experiences discomfort.

Over this and the next few chapters we'll delve into the science of how to regulate emotion at a physiological level, only later coming to the cognitive component of thoughts, beliefs, interpretations and analysis. We must first work with the right-brain implicit emotional processing because it is in this context that the intuition and empathy required for connection and secure relationship bonds are created. Also, because it is in the right hemisphere, with its connections to the body and the nervous system, that the primary control circuits for emotional regulation are located. Importantly, you can't read, think or behave your way to this state of connection born out of open presence and emotional regulation. You must learn how to feel it and work with it in your body because, as Damasio so beautifully puts it, 'Emotions play out in the theatre of the body.'

WHAT IS EMOTIONAL REGULATION ABOUT?

Imagine you're attending a performance by an orchestra. The members of the orchestra have been summoned to play their instruments but have been given no guidance on what they are to perform. The conductor raises his hand and they begin to play, producing a cacophony of notes clanging together with little cohesion, drowning each other out, ending in chaos. The conductor is rendered useless because he does not know what each person is playing and can't bring them to a point of harmony.

But think for a moment about the sound an orchestra can produce when they play the same piece; various instruments resonating in tone and rhythm whilst playing entirely different notes, the complexity of such a composition made almost simple because of the synchronicity between players. An emotionally regulated response involves the brain–mind–body emotion circuits working together in harmony, just like the players in an orchestra. Just as being dysregulated can bring about fragmentation and chaos. But what exactly is emotional regulation and how does this elaborate process of regulation play out in humans?

How do you regulate your emotions?

When you are faced with difficult or challenging situations, do you approach them with a reasonable level of confidence in your ability to cope? Or do you become easily overwhelmed, either expressing emotion inappropriately or suppressing it in the hope it goes away? Are you comfortable feeling your full range of feelings even when they are intense or negative, or do you prefer to block them with being busy, pleasure, denial, and other mechanisms designed to keep them out of your awareness? Are you able to acknowledge, tolerate and reflect on difficult, intense emotions in yourself and others – emotions such as anger, sadness, shame, disappointment and feelings of vulnerability – without rushing in to try and make things okay or go away?

You may not know how to answer these questions but a clue as to whether you habitually suppress emotion or whether you allow it to emerge lies in how wide a range of emotions, both positive and negative, you regularly experience. If you notice yourself being moved by emotions but are not destabilised by your feelings at a bodily level (i.e. feeling your heart warm and expand when you look into your child's or partner's eyes, feeling butterflies in your gut, feeling your chest and throat tighten in annoyance), you have the basic foundation in place for emotional regulation. The answers to these questions give you a sense of your ability to regulate emotions. But what exactly is it?

Emotional regulation describes your ability to respond constructively and flexibly to events with an appropriate emotion, at a level of intensity you can cope with and tolerate, and that is matched with the demands of the situation.

Stages of Emotional Regulation

Effective emotional regulation involves various stages and strategies. I have presented the elements that are most critical to parenting here:

1. Noticing and accepting emotions:
    This involves awareness of emotions and feelings in our bodies and minds, without judgement or the desire to switch them off or act on them immediately. It is about not reacting to the initial emotion with a secondary emotion such as anxiety about what we are feeling. It is the opposite state to avoidance, which involves trying to avoid, deny or suppress the emotion, which research shows to be unhelpful, and which in some cases, can amplify the negative emotion.

2. Soothing emotions:
    This entails the ability to calm or innervate the physiological sensations of our inner emotional states, for example lowering our heart rate and relaxing muscle tension, often involving self-soothing and compassion. This can occur simultaneously with stage one.

3. Reflecting on your emotions:
    This requires challenging distorted thinking patterns that produce or maintain unhelpful emotions (for example self-criticism, catastrophic thinking and general thinking patterns that are not based on factual evidence or logic). This also involves problem-solving to address the possible causes of unproductive emotions. I will present strategies to help promote healthy reflection in later chapters.

4. Taking action to deal with the emotion:
   This may avoid distraction, seeking help and comfort from others,
   assertively sharing our feelings with the people whose actions
   contribute to them, making a plan to manage a deadline, apologising to
   someone and any other action designed to regulate the emotion. Note
   that distraction may be helpful but only if it is based on accepting and
   noticing what you are feeling, rather than trying to avoid it.

How well we are able to handle our own emotional reactions depends
on whether we have developed the brain circuits that underpin healthy
emotional self-regulation, much of which we now know is actively
shaped in the first three years of our lives based on the type and quality of
mother–child interaction we received (please note I use the term mother
because it is most frequently mothers who have the most interaction
with their babies in the earliest stages). Most of us will have a broadly
consistent pattern relating to how we regulate emotions, especially in the
context of our relationships with others, but it is important to note that
this can be affected by difficult life events and also temporary states such
as tiredness and stress.

I can't begin to overstate how essential effective emotional regulation
is to our relationships, achievements, resilience and wellbeing, both for
parents and for children. It's worth taking a few moments to understand
what balanced emotional regulation is all about before we look at why it
matters so much to parenting. As you read through the next few chapters,
if you find yourself reflecting negatively on your own ability to regulate
emotions, please try to keep a sense of compassion for yourself and in
turn for your own parents. I find it helpful to remember the mantra of
compassion, put forward by the psychologist Paul Gilbert, who has spent
decades researching the science of compassion. As he puts it, we are
all doing the best we can with the brains we have at this point in time,
shaped by early experiences that we didn't choose, something that would
have affected our parents just as much as they did us.

Do also bear in mind that because of neuroplasticity, we are able to change and reshape some of these circuits in both ourselves and our children. I can say, hand on heart, that I have gone from having relatively unstable self-regulation skills due to early childhood trauma, to being capable of calm, warm and balanced parenting, of course not all the time but often enough for it to matter. So please don't allow your reading of what follows to create layers of guilt, anxiety, self-criticism or denial, but try to accept what we know to be true, based on extensive research in the field of neuroscience and parent–child attachment, and make a commitment to change if you feel that is best for you and your children.

An exercise to start to build your emotional regulation skills (eight minutes)

This exercise will help you to develop the skills of noticing your emotions and soothing them. Before you start, read through this exercise so you know what you need to do. You might also like to listen to my audio recordings on SoundCloud. Some of these instructions might sound strange to you but try to keep an open mind and do the best you can with it. I have based these exercises on the science I present to you in this book and on techniques put forward by respected scientists and practitioners in the field. Most of my breathing exercises will ask that you focus on the space behind your forehead as you inhale; what I mean by this is take your full attention to that space as best you can. Please inhale and exhale through your nose.

Start by finding a quiet space where you are unlikely to get distracted or disturbed. Set an alarm for eight minutes. Sit with your body in an upright and comfortable position, ideally with your arms and legs uncrossed. You might like to close your eyes so you can really focus on your body. Take a few moments to tune in to yourself and your inner sensations as you breathe in and out, whatever they might be.

1.   Relax – Take a deep inhale up into the space behind the middle of your forehead and breathe out into your body; let your body soften and relax as you exhale. As you breathe in, allow your belly and upper back to expand with the breath and exhale deeply in whatever way

feels comfortable to you. Try not to tighten your body because you're trying too hard. Repeat this up to five times or more.

2. Tune in – Now continue to inhale up into the middle of your forehead but exhale out into your heart and the space around it. Take your full attention to the area around your heart and the sensations there. Try to notice, without any judgement, the sensations you feel around your heart. Repeat this for three to five minutes or longer.

3. Notice – Widen your attention to your belly and gut area and notice whether you feel relaxed or tense. Exhale into the whole of your upper body, so that it feels as though your breath is almost travelling down from the space behind your forehead into your heart and gut.

4. Soften, slow down and soothe your inner state – see if you can use your breath to soften and relax your body, from your face all the way down to your gut. Let your exhale into the heart and belly area slow down and lengthen almost until there is no breath left. Please try not to tense up; this is a very gentle softening of your muscle tension along with an intentional slowing-down of your breath. See if you can allow whatever sensations are arising to just be, without trying to shut anything down, or react to anything, or even judge and label what is happening. Remember emotions are like waves that rise and fall – just let them pass.

5. Try to exhale a sense of kindness and warmth into the area around your heart and use this to soothe and calm your inner state. Notice how this feels and adapt the exercise according to what you sense you might need.

Try to do this exercise regularly, because as you know, change happens through neuroplasticity, which is based on repetition. The more often you do this, and other exercises in this book, the easier they will become and the more you will come to enjoy and benefit from them.

Chapter Five: Key points

- When we are able to regulate our emotions effectively we can lend our own brains to coregulate our children when they are in a dysregulated state. Children do not develop the capacity for sophisticated emotional self-regulation until they are older and can easily become overwhelmed, even in their teenage years.

- Embedded in our nervous systems is a rapid, non-conscious ability to read the inner emotional state of another person. Emotions are contagious between people, so your emotional state will have an impact on the emotional state of your child. Children are adept at reading non-verbal signals that signal the emotional responsiveness of a caregiver to them.

- Emotions are shifts in your inner physiological state in response to an internal or external stimulus. This could be a thought you have generated in your own mind, or it could be the non-conscious sensing of a change in someone's facial expressions or tone of voice. These shifts in your body and brain prepare you to respond to the situation, person or thing that has triggered the emotion.

- There are three components to an emotional experience: (1) the physiological sensations in the body; (2) the feeling (interpreting the sensations in the brain and labelling them); and (3) the thoughts, images and mental processing that accompany it.

- Emotional processing occurs mostly in the right hemisphere of the brain, where information about physiological changes in the body (heart rate, breathing, muscle tension) is first transmitted.

- Emotional regulation describes your ability to respond constructively and flexibly to events with an appropriate emotion, at a level of intensity you can cope with and tolerate, and that is suited to the demands of the situation.

- Effective emotional regulation involves four elements: (1) noticing and accepting the physiological sensations of emotions; (2) soothing and calming them in the body; (3) managing or challenging our thoughts about them; and (4) taking constructive action. Emotion regulation strategies based on denial, suppression and other forms of active avoidance tend to be less helpful and can in some instances amplify the negative emotion. Acceptance of the emotion without struggling against it is generally a healthy way to build emotional resilience.

• How well we regulate our emotions depends on whether we have developed the brain–body circuits that underpin effective regulation. With commitment and regular practice, it is possible to improve the way in which we regulate our emotions over time.

CHAPTER SIX

Emotional regulation and the window of tolerance

Healthy emotional regulation involves being able to authentically sense, tolerate and soothe your emotional reactions in a way that is helpful to the situation you are in. In the first instance, it is a physiological process because you need to be able to handle your bodily sensations and then manage what you do with them. When you're regulating an emotion effectively, you are able to experience the sensations that arise in your body without becoming overwhelmed by the intensity of these internal shifts. People who are able to regulate their emotions effectively trust themselves to notice, manage and respond appropriately to their emotions. Not only that, but they can differentiate their inner emotions from those of others around them, something young children struggle with. They can recognise and tolerate multiple emotions without immediately reacting to them. And most importantly, they can turn to others to help them regulate their feelings (in appropriate ways!) because they have learnt, from their early relationships, that their feelings will be respected, validated and soothed by the people whom they trust.

When we are regulating emotions effectively, we are at our best in terms of how we pay attention, our resilience and our adaptability. In this state, as Daniel Hill puts it, 'we are alert and all our psychological resources are available'[30]. Being able to remain mindful of how we are reacting, being present, and focused on the issue at hand, however uncomfortable we may feel, is central to this and gives us a sense of confidence and resilience in the face of stress. This type of resilience is different to the pumped-up positivity that stems from unrealistic or overly optimistic thinking, either about a situation or about ourselves.

When we are unable to self-regulate effectively, we tend either to feel too much or too little, mostly because we find emotions uncomfortable and we don't know how to manage them in a way that allows us to return to our state of natural equilibrium. It is important to bear in mind that *it is often not the trigger situations that we are afraid of but the discomfort that our emotional reactions to them may involve.* We become afraid of our emotions and the uncomfortable sensations that accompany them because they make us feel vulnerable or 'out of control'. Anxiety, for example, because it considerably raises the heart rate, can feel so disturbing that many people experience a secondary layer of anxiety about feeling anxious.

HEALTHY EMOTIONAL REGULATION REQUIRES BODILY AWARENESS

Emotional regulation is primarily a right-brain-mediated process that requires bodily awareness. Daniel Siegel describes this ability as 'interoception' – the ability to be aware of the interior of your body and the sensations you experience in different areas. I don't just mean pain, itchiness and those sorts of sensations but shifts in the activity of your nervous system and the subtle signals emanating from your heart and gut area. There are tiny receptors distributed throughout the organs, skin, muscles, bones and other parts of our bodies that gather information on the current state of those body parts and send this information up to the brain. Some people are more readily able to perceive these sensations compared with others, who may, in extreme cases, have very little understanding of what they are feeling on the inside. Whether or not you do will depend on how well-functioning certain brains parts are, including the insula, the part of your brain that can map physical sensations and recognise where they emanate from.

To learn how to regulate emotions effectively, it is helpful to focus on these four domains:

- your heart rate and whether it is getting faster or slowing down
- your breathing and whether it is deep and slow or quick and shallow
- muscle tension in various parts of your body, in particular your facial muscles, the area around your heart and chest, your neck and also your gut
- brain–body chemicals such as neurotransmitters that lead to various sensations in different parts of our bodies

CONNECTED PARENTING HAPPENS WHEN YOU'RE IN YOUR WINDOW OF TOLERANCE

An excellent way to understand how emotional regulation works is through the Window of Tolerance Model[31, 32].

We all have a naturally fluctuating level of nervous-system activation during the day depending on what we are engaged in or what we need to cope with at the time. We might be active and energised or relaxed and calm. Your window of tolerance represents the range of physiological (bodily) activation that you are able to tolerate without becoming dysregulated. When you are within your window of tolerance for an emotion, you are in a state that is *calm yet alert* and you are able to experience various emotions at a range of intensities without becoming overwhelmed by them. You might experience intense states of an emotion that take you to the outer edges of your window of tolerance without becoming dysregulated. For example, you might feel very frustrated when your child is dawdling with his shoelaces and you're already running late, without it tipping over into anger, or you may even feel intense irritation without becoming preoccupied, lashing out or raising your voice.

When you are within your window of tolerance for an emotion, you are able to authentically 'feel and deal' with the emotion with your mental faculties intact. You are alert, present in the moment, and able to fully take in what is happening; your mind is not jumping around and your heart rate is neither too fast nor too slow. In this state, the various brain regions involved in emotional regulation are working together in a connected way, sharing information and communicating so that you can respond flexibly to whatever is happening. You are able to differentiate your emotions from those of others and keep a sense of perspective. Your head feels reasonably clear and you can process what is going on around you, all the while remaining open to new information and possibilities. Although you might not think it, this hinges heavily on how well you are able to regulate your heart rate and breathing, because when your bodily reactions are regulated, your 'rational' thinking brain can stay switched on.

On the other hand, whether or not you notice this, when you feel overwhelmed by the sensations of the emotional response in your body along with its mental accompaniments (thoughts, interpretations, judgements), you move out of your window of tolerance and become mindless and preoccupied, reacting to things in a state of 'autopilot', rather than flexibly and consciously adapting your responses to best fit the situation you are in. You either become hypervigilant towards possible threats, or you become somewhat fuzzy and preoccupied, unfocused and distant, as though you are not fully present and open to seeing things clearly. You might find yourself talking quickly and extensively without really looking into the eyes of the person you are speaking to, or you might find yourself wanting to avoid feeling and dealing with the situation that is provoking them. You might become preoccupied with fixing or correcting the situation or person who contributed to the emotion, so that you can return to a state of inner comfort. Or, on the other hand, you may find yourself feeling very little emotion, almost detached and numb, wanting to distance yourself from people or the situation in order to feel safe and 'together'.

MOVING OUT OF YOUR WINDOW OF TOLERANCE INTO HYPERAROUSAL OR HYPOAROUSAL

When your level of physiological activation, in the form of changes to your heart rate, breathing, muscle tension and chemicals, exceeds your ability to cope, you move from a state where your brain, mind and body are working together and you are at your best in terms of focus and resilience, to a dysregulated state where the various circuits stop working as a team and you become stuck in a box. This leaves you unable to think, respond or even learn constructively. Because we are primed to maintain a level of physiological equilibrium, moving into a dysregulated emotional state feels uncomfortable, often deeply uncomfortable for children, and may create a drive either to release the emotion in some way (sometimes not in a constructive way), or to suppress it.

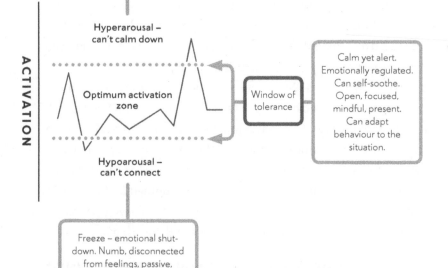

The Window of Tolerance.
Adapted from Pat Ogden; Fisher; Sensorimotor psychotherapy

As you can see from the diagram above, when you move out of your window of tolerance, your level of nervous-system activation may go up in **hyperarousal (high)** or down in **hypoarousal (low)**. With hyperarousal your physiological reactions become heightened beyond what might be useful in that situation, in other words, your bodily sensations, such as breathing, heart rate and muscle tension, become raised and magnified, to the point where they may feel very uncomfortable. From an evolutionary point of view, this elevated state of activation was adaptive because it enabled us to either attack a predator or run from it. However, many of the situations for which we now experience a state of hyperarousal do not necessitate such an intense level of physiological activation.

On the other hand, the state of hypoarousal occurs when we become switched off and numb to our emotions, akin to freezing. In this state, the heart rate and breathing slow down and the sensations in the body become dulled and anaesthetised to a point where we don't feel enough to deal with the situation constructively. Do note that all this can happen within seconds and is not something many of us are aware of in the moment. We can also vacillate between states of hyperarousal (up) and hypoarousal (low), depending on the trigger situation, our tolerance for that particular emotion and our general state at the time. Please note that this model applies to both positive and negative emotions, though negative emotions typically evoke a stronger bodily response and are therefore more likely to become dysregulated.

What does hyperarousal look like?

In simple terms, hyperarousal represents an escalation of the emotional response that usually results in a desire to take some form of action to alleviate the discomfort. Anger, panic, anxiety, agitation, excitability, anticipation, distress and embarrassment may all result in a state of hyperarousal. When the trigger is sufficiently threatening, your 'fight or flight' response will take over, which leads to the desire either to fight back verbally or physically in self-defence, or to turn away from something threatening. You might struggle to relax, experience sleep disruption and find yourself feeling stuck in a defensive, agitated state. You may experience a heightened awareness of potential threats, creating hypervigilance. Some people relieve the built-up emotional energy by taking action, for example, talking excessively, shouting, moving their bodies and so on. Others relieve it by turning away from people and distancing themselves from the perceived trigger of the emotional response.

In the state of hyperarousal, we sometimes use people around us as 'receivers' for our emotions when we release them, perhaps by unburdening ourselves to them, or by venting or lashing out to rid ourselves of the uncomfortable build-up in our bodies. But we can also move in the other direction and distance ourselves from people, in an attempt to 'contain' the emotions and put them in a box. In either case, we go into an 'autopilot' mode that makes us hasty, impulsive and unable to sense intuitively our impact on the person we are communicating with.

As parents, a state of hyperarousal often reveals itself in angry shouting, telling our children off harshly, or even asking them to 'go away' – I'm sure every one of us has at some point felt how difficult it can be to stop this type of response from escalating into something we later regret! If something difficult occurs and you momentarily tip out of your window of tolerance, your child will sense this and begin unconsciously to mirror your emotional response (unless she has already learnt to shut off from difficult emotions). Your child's sense of psychological safety plummets and he or she begins to feel anxious, afraid or uncomfortable. If your emotional response is not too intense and you are only just out of your window, your child might experience only a small degree of discomfort being around you. But if your emotional response is escalating, your child will sense this, move out of his own window of tolerance and respond by either venting his own emotions or shutting down to protect himself from the effect of your emotions on him.

If you have healthy regulation circuits, and you are able to mostly stay within your window of tolerance, or if you have tipped out of it, but return to it in a reasonable timeframe, you will quickly calm down, at which point your child will feel emotionally safe and relieved. If you're able to then connect with your child and repair the situation with warmth, the episode is not likely to leave a lasting imprint. However, if you experience an intense uncontrollable episode of emotion, and you are unable to reflect on it, calm down quickly enough or repair it with your child by restoring connection and/or talking about it, it will become embedded in your child's implicit (non-verbal) memory system. Over repeated episodes, this will leave a sense, deep below the level of consciousness, that emotions are frightening experiences and your child cannot trust people to help him or her manage them.

Hypoarousal makes connection difficult

Hypoarousal involves a lowering of the heart rate and physiological activity and a numbing of the emotional response, resulting in the sensation of stillness, collapsing or shutting down. This tends to occur when we feel we are powerless to prevent or stop the cascade of deeply uncomfortable emotions that arise in situations we feel helpless to resist or control. This may result from a sense of perceived hopelessness after a period of hyperarousal that cannot be managed or alleviated. In the extreme, it causes a 'freeze' reaction where our heart rates drop very low, we may faint or vomit, and we cannot feel much at all, akin to 'playing dead'.

This is an ancient survival mechanism that we share with reptiles and its adaptive value is that we grossly reduce metabolic demands and energy usage by slowing down our physiological processes. However, once again, in current times, this form of self-protection is not always useful, though it can be if, for example, we are under attack and have no ability to fight back or run away from the source of threat. This state of hypoarousal, in the extreme, represents what psychologists call dissociation and it reveals itself in a 'turning away' from connection, emotions, our own bodies and their sensations. Boredom, sadness and shame are examples of feelings associated with either mild or moderate hypoarousal.

Melancholy, hopelessness and suicidal feelings may be at the more extreme end.

Those of us who are prone to hypoarousal haven't learnt how to soothe uncomfortable sensations and emotions so it becomes a far safer strategy to shut down and deny feelings altogether. A habitual tendency towards hypoarousal might emerge in people who have experienced trauma, or whose emotions have not been understood and handled with delicacy and compassion. Children who are regularly shamed, criticised, ignored or neglected, especially in relation to their emotions, may develop a tendency over time to shut down inside and dissociate from bodily sensations. Because they haven't learnt to self-soothe emotions, they gradually become afraid of them.

In the state of hypoarousal, you may find yourself feeling detached and emotionally disconnected from your feelings or those of your children, your eyes might glaze over or look expressionless, you feel a bit numb inside, and you struggle to experience a sense of 'felt empathy' for others. You might find it hard to really care about things, tending towards indifference or listlessness (bear in mind I'm describing this as a temporary state rather than a trait). You may even feel tired, low in energy and sluggish.

In parenting terms, hypoarousal or dissociation could manifest as a lack of eye contact, not understanding or feeling interested in your children's emotional needs and responses, not feeling pleasure and joy in being around them, or simply feeling emotionally detached and disconnected from them. In milder cases, it may appear as a lack of warmth and emotional bonding, as may be the case when a parent suffers from depression, and in the extreme, parental neglect. By this, I'm referring not only to looking after our children's physical needs such as providing food and taking them to school, but also being out of tune with the inner emotional life of your child.

We all have habitual patterns of emotional regulation

Some of us have a wide window of tolerance that empowers us with the ability to tolerate more intense states of emotion without becoming dysregulated. Others, like children, may have a narrow window, with even mild to moderate states of certain emotions nudging them out. There is a natural hierarchy in the pattern of emotional regulation; typically, we move towards hyperarousal (high; fight or flight), and if that is intolerable or fails to relieve us of our discomfort, we might switch towards hypoarousal (low; freeze). How quickly you move out of your window of tolerance depends on the strength of your emotional regulation circuits, the emotion in hand, the situation, and your threshold for the emotion at the time, something that might vary depending on your general state, for example, how stressed or tired you are. In some cases, people habitually resort to one or the other because they have learnt these strategies in childhood and their brain circuits have become wired up accordingly.

I have worked with many clients who tend to suppress emotions and put on an 'it's all fine' front but if the trigger situation is not resolved or removed, as it frequently isn't given they are not aware of their emotional states and how to soothe them or how to reach out for help, they experience huge outbursts that take them by surprise. This is because they have not been able to recognise and tolerate their emotional responses and have instead taken to suppressing or denying them. But this strategy doesn't work very well in the longer term because the chemicals that are released when we are stressed or experience other emotions don't just go away; they continue to build out of our awareness so the tipping point for an emotional outburst is gradually lowered. Some people who do have an awareness of their emotions in a state of hyperarousal, but have not been able to resolve their distress either independently or through reaching out to others, may later become disengaged, switched off and disconnected as a result of their growing sense of futility. This is when they may find themselves feeling depressed or desperate, and switching into a state of prolonged hypoarousal.

Features of healthy emotional regulation

Here are some reminders of what effective emotional regulation involves:

- The emotion you experience is appropriate to the situation in hand.
- You can notice and tolerate the intensity of the emotion without tipping into hyperarousal or hypoarousal, unless you are facing a serious threat in which case it is entirely natural for you to experience the states of fight, flight or freeze.
- When you do tip outside your window of tolerance, which is natural and will happen at times, you can return to it quickly and relatively easily. This is where calming and soothing emotions plays a part. Though in states of hypoarousal without a threat, for example boredom, sometimes energising rather than soothing is required.
- You can think and reflect on your emotions in the moment or soon after.
- You can seek help or support in dealing with difficult emotions in ways that are appropriate and reasonable.

AVOIDANCE AND COMPENSATION: UNHELPFUL COPING STRATEGIES

When we are uncomfortable with intense emotions, we often develop unhealthy coping strategies to help us, some of which involve a state of avoidance or compensation. You may find yourself avoiding certain situations that you know trigger emotions that you struggle to tolerate, or when the emotion is triggered in spite of your efforts, you may (non-consciously) numb yourself to the sensations of the emotion. Both of these represent a form of avoidance.

On the other hand, you may find yourself compensating for certain emotions that you don't like to acknowledge, perhaps through left-brain-driven mechanisms of control such as rigid routines, exercise and other activities that either boost your sense of self-esteem or your ability to avoid feeling emotions that make you uncomfortable. Many people who consume drugs or alcohol at extreme levels may be seeking ways to regulate their emotions, sometimes attempting to feel alive in the face of habitual suppression of emotion, or seeking to numb the emotions by drowning them in pleasure (sadly it doesn't work). Some parents may inadvertently use their children to help keep certain emotions about themselves at arm's length, for example gaining vicarious satisfaction from their children's experiences to mask their fears that they haven't or won't achieve enough themselves, or becoming too attached to their children as a defence against the fear of rejection or abandonment. Parenting through rigid rules is often a form of compensation because these parents need control to avoid unpredictability, which they find destabilising and threatening.

HOW WELL YOU ARE ABLE TO REGULATE YOUR EMOTIONS AFFECTS YOUR CHILDREN

Healthy emotional regulation requires bodily awareness via strong, well-connected brain–body circuits, particularly in the right hemisphere. It also requires the ability to think and reflect on our own emotional and mental states, something that is made possible by the middle pre-frontal cortex area of the brain, just behind our foreheads. When we move out of our windows of tolerance on either side, we lose our mental and emotional flexibility and revert to either chaotic or rigid responses that aren't matched with the unique needs of the particular situation we are in. This is a transient state that may recur frequently in the course of one interaction with someone. For some it may be momentary and for others it may persist for periods of time.

In both these states, your capacity for empathy and problem-solving are off kilter and you become preoccupied, mindless, self-focused or detached. States of hyper- and hypoarousal change the way we are 'present' in the moment and also how we engage with the people around us. It is in states such as these, when we are on a sort of emotional 'autopilot' system, that our children become vulnerable. They become so because when we are unavailable to notice their emotions because we are preoccupied with our own internal state, they cannot bring themselves to a state of inner calm and equilibrium. Because we are not able to feel much empathy when we are dysregulated, we tend to see their behaviour in a negative light, and we are cut off from accessing more plausible or even compassionate explanations for their behaviour. This leads to ruptures in the level of connection between children and parents.

Your emotions are contagious so your children feel them too

We mirror each other's emotions without recognising we are doing so. Because our brains and bodies are designed to synchronise with each other through the detection of shifts in our internal states, through the tiny micro-movements in our facial expressions and changes to our tone of voice and muscle tension, emotions become contagious. When you feel tense inside, your facial and bodily muscles will reflect this, your tone of voice will subtly shift and your child will sense this, right brain to right brain, with no awareness that this is occurring. Your child, in sensing your emotional state, will have a reaction to it so that his heart rate and tension levels rise or lower in response to yours. Children implicitly take in whether you are relaxed or not around them. This is how implicit communication works: it occurs in a matter of milliseconds and is independent of verbal language.

This language of the body is faster and more powerful than the verbal language system, which came much later in the story of human evolution. This emotional contagion or synchronising of emotions occurs too quickly for conscious awareness and management, though you may become conscious that the emotional tone between you and your child has shifted after it occurs. When children frequently experience dysregulated emotions through their interactions with you, they come to develop a stable pattern of emotional regulation that is designed to help them feel safer in their interactions with you. That could include shutting themselves off to you or clinging to you and magnifying their reactions in an attempt to seek comfort. Rather tryingly, when you are at your most dysregulated, they will mirror this back to you, making it even harder for you to bring yourself back to your window of tolerance, much like trying to swim against a tide.

Children mirror back to us their reactions to our own internal state

When you are stuck in a pattern of dysregulation, which you frequently may not notice until you are quite far down this path, this will be reflected back to you in the behaviour of your children. Inevitably, when I notice my children are becoming argumentative, critical, whiny, or just a bit difficult, I take a hard look at myself and discover that sure enough, there has been something looming in my schedule that has made me preoccupied and mindless with them. Perhaps I haven't been getting enough sleep, or have been trying too hard to fit too many things in. Sometimes it has been about just one event, or other times about a build-up of things I'm anticipating in the weeks ahead. These occurrences may not be things I consciously consider to be stressful at all, and that I appear to be coping with well, but my body reveals what my mind conceals, and it is this hidden language that speaks to my children.

It declares itself in small things like the tone and pace of my speech, whether I respond to their conversation in autopilot mode or whether I fully engage with them by looking into their eyes and slowing down. Often it announces itself in more damaging ways like a shift in how critical or judgemental I become of the things they do and say. It changes how I see their behaviour and whether I generate a kind and benevolent explanation or a negative one. I may not express it this way to them, but it feels like the difference between 'You sometimes get things wrong and that's okay' and 'What is wrong with you and why is parenting you so hard? Why can't you just be different, easier, go away and give me space ...' They pick up on this undertone of stress and resentment, however subtle, and it makes them feel disconnected, defensive, ashamed or sad. When this happens, I need to pay attention to relaxing and cultivating the state of integration between brain, mind and body that my children need to feel safe and happy around me. I know when I am at my best because that is when parenting becomes more joyful and effortless. The children become kinder, softer, more compliant and eager to please.

Children need to learn how to manage emotions to develop a healthy sense of themselves

Regardless of how they appear on the surface, people who tend towards suppression of emotions, either by ignoring their emotions when in a state of hyperarousal or through hypoarousal and dissociation, don't have a healthy sense of themselves because they cannot feel or process their own emotions and therefore can't really know themselves at a deep, intuitive level. They may rely on left-brain verbal and analytical processing of emotions, which as you might recall is compromised because the left hemisphere has no real felt sense of connection with the body and the world around us.

When emotions are not felt, recognised and validated, we begin to rely on extraneous factors for our self-esteem or to know 'who we are'. We experience ourselves only through what we do and think, rather than how we feel. We often don't know what we really value, what makes us truly happy, or what we need to nourish ourselves and live a balanced life. Because those prone to emotional suppression or hypoarousal don't recognise their feelings, or know what they mean, they are left to navigate through life without an inner compass to guide them. The scary part is that they often don't know what's missing. Emotions organise our sense of self so it is really important that we learn how to recognise and deal with them in our children.

HOW DO YOUNG CHILDREN REGULATE EMOTIONS?

Children are not born with the ability to regulate their emotions effectively because the brain circuits responsible for regulation tend to develop over the first few years of life. Very young children have only a small window of tolerance for their emotions and can only respond to their emotions in the most rudimentary way, e.g. sucking their fingers, averting their gaze, or crying and screaming to signal they need a caregiver to soothe their distress for them. Children also struggle to differentiate their emotions from the emotions of others and initially do not know the difference between their own feelings and those of people around them. This makes it harder for them to resist being triggered into an emotional response by the unfolding emotions of those around them. They are easily overwhelmed and are programmed to release emotion via hyperarousal in the first instance, unless the situation is so excessively threatening to them as to elicit the 'freeze' state of hypoarousal. Emotions on the shame/vulnerability spectrum can be especially uncomfortable to handle, as is any form of perceived rejection. We know from brain imaging studies that rejection activates the actual centre for pain in the brain so is it any wonder that without having developed the ability to calm such difficult sensations, children try to protect themselves by either suppressing them or lashing out?

When children become dysregulated and move into hyperarousal, they are even less capable of inhibition and self-control than us adults so name-calling, screaming, throwing things or crying out of anger are all examples of how they might attempt to handle their emotions without the requisite brain circuitry for regulation. Even when the emotion is positive, as in the case of excitement, some children will struggle to tolerate the intensity and can rapidly flip into the fight or flight mode.

When they move into moderate or extreme hypoarousal (low), they seem frozen in their tracks, unresponsive to what you might be saying, with flat facial expressions and a fixed gaze. Young children often find it hard to tolerate states of hypoarousal such as boredom, particularly when there is no choice to move around or engage with something, and successful parental coregulation in this instance involves noticing and providing them with stimulation to bring them back to their window of tolerance. This needn't be done immediately but if they are becoming very dysregulated from a lack of stimulation an attuned parent will usually automatically engage with the child.

It is important to remember that children are not trying to vex you, annoy you or embarrass you (it's really not about you at all), or to be difficult or behave badly, but are usually attempting to deal with their emotions in the only way they know how. This is why your own ability to remain regulated in the midst of all the emotions they experience (and trigger in you) is so vital to the parenting relationship. Your ability to regulate their emotions successfully, invisible as this process may feel to you, is the crux of their sense of psychological safety.

Some children, in response to poor parental regulation of emotions, may shut down. Others learn that to get your feelings noticed you need to let them out and dramatise them; you must attract attention by any means and hope that by making all this noise, your parents may come to realise what you are feeling and help you to calm your distress. Whichever way the child goes, the point is that they become overwhelmed by inner emotional states and rather than being able to calm themselves back into their window of tolerance, they flip into hyperarousal or hypoarousal, which, over time, becomes their habitual strategy to manage emotion.

WHY WE NEED TO STOP USING OUR WORDS

The brain hemispheres play a critical part in the way in which emotional regulation patterns develop. The left hemisphere, with its tendency towards verbal and systematic analysis, tends to develop in the second year of life [33] but prior to that children are dependent on the non-verbal emotion-processing capabilities of the right brain. This is important because we have a preoccupation with encouraging children to 'use their words' to express emotion but knowing what we now do about the importance of the right hemisphere to the processing of emotions, I think it makes little sense to coax children to abandon the emotion as an embodied experience and to disconnect from it prematurely through putting it into words. We do this because it makes it easier for us, with our left-hemisphere dominance, to understand and deal with what they are experiencing. In the next section I describe how to let children feel their emotions fully before jumping in to find ways to distract them, or even to fix their problems and find solutions.

Even in grown-ups with better developed left-hemisphere functioning, the rational verbal analysis of the left can only help us regulate mild to moderate emotional states. Regardless of how competent a left hemisphere we may possess, when we move out of our windows of tolerance with an intense emotion, it is the right hemisphere that is in charge and therefore it is really essential that we understand how to address emotion in a non-verbal way and allow this process to unfold in the way that nature has designed. To reverse this trend towards cool, rational analysis of emotion and independent self-management, we need to allow our children to feel and tolerate emotions at a bodily level first. Putting things into words is useful and important as a second stage in the management of emotion but only once the intensity has lowered and your child has felt soothed at a physical level. This is how children learn to tolerate the sensation of emotion and feel a real sense of connection with themselves.

This can be done beautifully with no talking at all. It requires silence, space and a calm, relaxed bodily state; we really don't have to rush in to respond to their emotional outbursts with words, analysis, comfort or anything in the first instance. If you are in a right-brain-mediated mode of open connection, this will happen of its own accord. When it does, it tends to unfold slowly enough that the child gradually learns to tolerate discomfort for short periods of time before you step in to soothe the emotion or remove the source of it. Your child needs to experience discomfort and you must facilitate this and manage your own reaction to it. I regret verbalising and addressing things too quickly in my early parenting years with my first child; I had so much empathy for him that it often became dysregulated and I would rush to soothe him before he could learn to tolerate the faintest frustration. Because of my high empathy and connection with him (but not enough differentiation), I often anticipated potential sources of distress for him and didn't allow the possibility of negative emotion to even occur.

HOW DOES EMOTIONAL REGULATION WORK IN PRACTICE?

Here's a simple example of emotional regulation between a child and an adult. Just last week, I came across a four-year-old child sitting in the hallway at school, silently crying and refusing to go to the car with his mother. She, naturally, was trying to cajole him to get a move on, using all the reassuring language she could think of. She seemed calm but her eyes were preoccupied – she had things on her mind, things to do, and he was getting in the way. His older sister was getting restless and no doubt their mother was aware that the whole incident could soon take a turn for the worse. The little boy wasn't making eye contact with his mother and didn't want to engage with her. He was determined just to sit there, as unmoveable as a rock. As I walked past him, I stopped and looked at him, and noticed he looked distant but also forlorn – he was clearly out of his window of tolerance and needed help regulating his emotions. His mother explained that he didn't want to leave school because he would miss it too much over the summer.

I stopped and crouched beside him, looked into his eyes, and said quietly, 'You're sad because you really love coming to school and you're going to miss it, aren't you?' I felt warmth and kindness towards him and it softened my tone of voice. He looked up at me slowly and nodded. His mother said loudly and cheerfully, 'It's fine because you'll be back before you know it!' She was trying so hard to be positive and light-hearted for him but because it didn't match what he was feeling, he didn't respond. I softened my tone of voice even more and said calmly, 'You really don't want to leave because you've had such a lovely time.' He looked into my eyes and said, 'Yes'.

I could see that he had stopped crying and was more present. At this point, he was moving back into his window of tolerance and was able to pay attention to what I was saying. I said, in a light-hearted voice, 'The next few months will fly by so fast and then you'll have a whole year at school again.' Without even replying, he jumped off the chair, held his mum's hand and simply walked away with her. He went from being dysregulated to regulated in no more than a few minutes; it was not really my words that helped him achieve this but my face and tone of voice. It was feeling understood whilst being with someone who was in a regulated emotional state that allowed him to make this shift in his own internal state. Of course, it was so much easier for me to feel this connection with him because he wasn't my child and I wasn't distracted by what it meant for me that he was crying and delaying leaving school, or how it looked to others. But it is possible to be like this with our own children too; not all the time, but at least enough for them to develop healthy emotional regulation skills over time.

HOW TO COREGULATE EMOTIONS WITH OUR CHILDREN EFFECTIVELY

Emotions are a fundamental and inherent part of being human. However uncomfortable, they are not something to be afraid of but rather a source of information to listen and respond to. They often tell us about what we value, what we need, who we are and what we fear. But do bear in mind that recognising an emotion and showing a regulated degree of empathy is very different to agreeing with the causes or consequences of it. Emotions can never be wrong or bad but the way in which we express them might be helpful or unhelpful to the situation at hand. Children need to understand that it is okay to feel their feelings in all their glory or otherwise, but they are not free simply to express them however they wish, with little consideration for the impact on others. Teaching our children to feel and express their emotions openly yet appropriately allows them to develop an authentic sense of themselves; to perceive their own needs, wants, experiences and values as valid and at times worthy of sharing with others.

You can have an influence on how your children learn to experience and express their emotions. There are two ways in which this transpires: firstly, what you feel inside at the level of your nervous system, which they sense through your body language, and secondly, what you say and do. We often minimise or belittle our children's emotions because we judge them according to standards that might apply to adults. We sometimes rush to dismiss their feelings because we haven't the patience to deal with them constructively at that point in time. Perhaps we ourselves don't understand emotions very well and are uncomfortable because our children's emotions trigger disagreeable feelings in us. Whatever the reason, responding empathetically to our children's feelings often doesn't come naturally.

Here are a few tips to bear in mind when your children are in the grip of negative emotions. Remember the three stages of noticing, soothing and reflecting.

Noticing, accepting and soothing:

First, just let your child feel his feelings in a safe way with you, whilst you use your own regulated state to soothe and calm your child back into his window of tolerance. For this to work, you need to be regulated – calm, relaxed, present, attentive, open. Remember this is a biological state so take a few slow breaths, focusing on lengthening your exhale and softening your face and the area around your heart.

- Don't automatically rush in to talk about the emotion, reassure them or even shut them down by providing a source of distraction (unless the situation really demands it). Examples of these tactics might be offering them the iPad, food or some other form of pleasure to divert them from their emotional state. If it is a very mild emotion, distraction techniques such as diverting their attention or using humour may be helpful but not if it encourages them to turn away from their feelings too quickly.
- Don't try to control or distort their emotional experiences by minimising or dismissing them. Saying things like, 'Come on, it's not that bad', or 'You know you're okay really', or 'You've been through this before and you were fine!' are examples of minimising their feelings. We also frequently dismiss our children's feelings, for example by saying things such as, 'Don't be silly, you're overreacting' or, in a belittling tone, 'What are you upset about? It's nothing!' or 'Go to your room and calm down.'
- Remind yourself that it's the most natural thing in the world for a human being to have an emotional response to something and whilst you might have a rational mind that can modify and control your emotions, your child has a very rudimentary 'off button' for his emotions and must feel very uncomfortable. He needs your help and kindness, not your judgement.

- More importantly, you don't need to react or worry if things seem out of control because emotions can be calmed via soothing, empathy and a bit of rational thinking. In the case of extreme demonstrations of emotion in very young children such as toddlers, sometimes they just need to run their course. You might need to work harder to avoid being triggered by a tantrum yourself though!
- First, focus on yourself and remember you are your child's proxy brain for emotional regulation. They cannot self-regulate effectively if they haven't developed the brain circuits for this yet and it is how well you coregulate them that helps these circuits to grow and strengthen.
- Take a few deep, slow breaths, relax your muscles and your face and try to be present in the moment. If you feel agitated or wound up, lengthen and soften your exhale so it is slow and gentle. This will calm your heart rate and bring you back into your window of tolerance. Try to generate a feeling of softness and warmth around your heart towards your child.
- Make eye contact and for a few moments just let your child feel his feelings in your presence without any pressure to hide, change or control how he feels.
- If it feels right, try to soothe your child by gently stroking, holding, cuddling or just being next to him in a relaxed and kind way. Your child will naturally gravitate towards you if you are genuinely calm and compassionate and your voice is soft and gentle. Try to convey a wordless sense to your child that it's okay to feel bad, it won't overwhelm him or you, and you're right here to help and comfort him. This is all done at a non-verbal level. Talking softly at this stage is okay but often unnecessary.

Reflecting:

Ideally once your child is back in his or her window of tolerance and you have allowed some time just to feel and be soothed, you can begin to coach your child to recognise and deal with emotions productively. If the soothing phase has gone well, there may sometimes be no need to verbalise, correct or teach anything, though helping your child learn how to label emotions and reframe situations can be calming and helpful to emotional regulation.

- When your child starts to calm down, you might want to ask some questions to understand what led to the emotional reaction. Try to keep an open mind. However well you think you know your child, you will bring your own biases and preconceived notions to this process so try to slow down and explore the situation from your child's perspective. You might hear things that trigger critical judgement or emotions in you, but hold back as much as you can from putting your own point of view across at this stage.

- Try to show empathy by reflecting what your child is feeling and saying. Try to really 'see' and 'hear' the underlying message and show your child that you understand it from his perspective. Where possible, you can label the emotion so your child develops a vocabulary for emotion. For example, in a light-hearted tone, 'You must have felt really annoyed when you waited patiently for your turn and your friend refused to give you the ball.' This works best when you are in a calm and regulated state and your voice is gentle and relaxed. If you are too distressed by your child's emotional state, and this becomes evident in your voice, your child won't calm down or open up as effectively.

- When your child is a little more serene and feels listened to, you may want to address the behaviour and expression of the emotion. An unthreatening way to do this, without triggering too much of a shame reaction in your child, is to ask questions so your child can work out the consequences for himself. For example, 'When you shouted at your friend and pushed him, how do you think he felt?' 'Do you think it made him want to listen to you and give you the ball, or did it make him angry?' 'What else could you have done?', 'What made it so hard for you to wait a few minutes for your turn?'

- Note that the tone of voice you use will determine whether the child perceives your questions as hostile and judgemental or curious and neutral. Abandon the questioning if you are becoming fixated with whether you child is answering them the 'right' way or not.

- Explain to your children that the way we speak to each other changes how we feel about each other in that moment. People generally listen better when they are not feeling angry or upset. Also encourage your child to reach for help in resolving situations that become heated or emotional. This may involve asking you, or other reliable grown-ups who may intervene on your child's behalf, for help.

Whether or not you decide to explore these techniques further, I want to emphasise one last time that heartfelt connection comes first, and most of the more prescriptive behaviours I've discussed with you in this and earlier chapters will flow almost naturally when the foundation is firmly in place. Over time, with regular practice and effort, we can all achieve this to a degree that is good enough. Let's not forget that, as in all things, balance is essential to parenting. Without it we get caught up in rigidity and are easily catapulted into a state of threat and defence. Let's keep at the forefront of our minds that we are aiming for 'good enough' and not parenting perfection.

You'll know when your parenting is good enough because you'll feel it; your children will gradually become more open and willing to listen to you and your family life will be calmer, though it will never be perfectly easy and pleasurable all the time. When there is real connection between you for even a reasonable amount of time, your children will tolerate a higher level of rupture in the relationship. When your children feel emotionally safe and trust you to regulate your own and their emotions reasonably well, you'll start to feel you are moving with each other rather than against each other, and ultimately you begin to enjoy and nourish each other in a truly reciprocal way.

An exercise to build and practise emotional regulation skills: learning to calm and soothe uncomfortable emotions

Please do this exercise after you have practised the exercise in the previous chapter a few times. As with most of my breathing exercises, I will ask that you inhale into the space behind your forehead; what I mean by this is take your full focus and attention to that space as best you can. When I talk about exhaling into your heart or belly, you are taking your full attention there and using the breath to soften, soothe and relax the area of focus.

Start by finding a quiet space where you are unlikely to get distracted or disturbed. Sit with your body in an upright and comfortable position, ideally with your arms and legs uncrossed. You might like to close your eyes so you can really focus on the interior of your body. Take a few moments to tune in to yourself and your inner sensations as you breathe in and out, whatever they might be.

1. **Recall** – Bring to mind a situation that has caused you to feel a negative emotion in the parenting context – on a scale of one to ten, with ten being the most awful you could feel, try to pick something around a six or seven.

2. **Experience** – Now let the emotions you felt during this episode unfold without trying to do anything about it at first – just notice the four markers of emotion in turn.

   • Heart rate – does it go up or down?
   • Breathing – can you notice any changes to your breath?
   • Muscle tension – pay attention to the muscles around your heart, chest, gut, face and throat. What do you notice? Are there areas of tightness, constriction, heaviness or lightness?
   • Chemicals/sensations – what do you feel and where do you feel it?

This may be very hard for some of you but do persist. Instead of tightening and tensing up, or judging and analysing the emotion, or even escaping into your thoughts, just let it be for a few moments. Remember, it is just the language of your body communicating with you and it won't overwhelm you because you know how to soothe it through managing your breath. Try not to react to it.

3. **Soothe and calm** – Now, once you've tolerated the emotion for a few moments, start to calm and soothe it using your breath. Trust that you can slow your heart rate down and release muscle tension by using your breath. Take a deep inhale up in to the space behind the middle of your forehead and breathe out into your body; let your body relax as you exhale. As you breathe in, allow your belly and upper back to expand with the breath and exhale deeply in whatever way feels comfortable to you. Try not to tighten your body because you're trying too hard. Repeat this up to five times or more.

4. **Tune in to your heart** – Now continue to inhale up into the middle of your forehead but exhale out into your heart and the space around it. Take your full attention to the area around your heart and the sensations there. Use your breath to soften and relax that area. Imagine your exhale is a form of kindness and compassion for your body and breathe this sense of gentleness into all those tense muscles around your chest. Slow down your exhale so it is longer than your inhale. Repeat this with your belly.

5. **Notice what's changing as you relax** – Focus your attention on the parts of your body that need to be soothed and calmed the most. Soften your face and uncrease all the tiny muscles there. Try to lift the corners of your mouth upwards just a tiny fraction into the beginning of a gentle smile. Exhale into the whole of your upper body, so that it feels as though your breath is almost travelling down from the space behind your forehead into your heart and gut. Notice any changes that have happened.

6. **Engage your mind** – Once the sensations of the emotion in the body have lessened, try silently telling yourself something calming as you exhale. You might like to say, 'It's okay, I can handle this' or, 'There are different ways to look at this' or anything that helps you reframe the situation and feel more competent in managing your reaction to it.

Repeat the exercise – Go through the exercise again and repeat until you feel reasonably calm when you bring to mind your trigger situation. It helps to use this exercise with trigger situations involving your children, as repeated practice will teach you to react less in real life.

Reflection: What has shaped your own style of regulating emotions?

Our patterns of emotional regulation become embedded in the fabric of our relationships and how we connect with the people in our lives. Can you identify the emotional regulation strategies your parents used with you and how tuned in they were to your emotional needs as a child? Were they aware of your feelings? Did they react too quickly or empathically to your distress or discomfort and try to protect, even overprotect you? Or did they attempt to toughen you up by dismissing emotions and encouraging you to do the same? Perhaps they were uncomfortable talking about things that were personal and emotional and didn't speak about feelings at all, focusing instead on the things you did rather than how you felt. Perhaps they were unaware of what you were feeling and you, because you have developed the habitual tendency to suppress emotions, think it was fine and all this is a lot of overanalytical psychobabble? Whatever you think, I hope you'll stay with me and learn about your own 'relationship style' as a parent in the next two chapters because they are all about 'attachment' and the characteristic styles we adopt in our relationships with people, and especially, our children.

Chapter Six: Key points

- Healthy emotional regulation involves being able to notice, accept, soothe, tolerate and manage your emotional responses in a way that is appropriate for the situation you are in. To do this successfully you need to be able to stomach your inner sensations without feeling overwhelmed or wanting to shut them down.
- Your window of tolerance represents the range of emotional intensity you can tolerate without becoming dysregulated.

- When you tip out of your window of tolerance, you move towards either hyperarousal (high) or hypoarousal (low).

- As parents, hyperarousal can manifest as shouting, agitation, anger, anxiety, extreme concern, crying, or wanting to get away from the child. In parenting terms, hypoarousal manifests itself in a low level of intimacy, connection and interest in your children, along with a desire to want to downregulate emotional expression.

- Children need to feel their emotions fully in order to develop a healthy and coherent sense of themselves, because knowing ourselves in an authentic way requires a sense of connection with the language of our bodies and our emotions.

- Children are prone to hyperarousal because they have underdeveloped self-regulation circuits and rely on us to regulate their inner state.

- Bear in mind the four stages of emotional regulation when dealing with emotional episodes with your children – noticing and accepting, soothing, reflecting and taking action. Allow your child to experience emotional states whilst you calmly coregulate him or her before you attempt to talk about the problem or situation.

- If your child is in an extreme state of fight or flight or in the middle of a tantrum, you may just need to give her some time to calm down. It is important to stay near her and regulate your own emotional responses so that you can remain in your window of tolerance. If you slow your own heart rate and generate an inner state of compassion, warmth and calm, your child will sense this and her physiological state will begin to synchronise with yours.

- Once your child is calm, is able to make eye contact with you and can respond to an attempt at dialogue, you might choose to talk through the situation and help her to understand what happened, why it happened and how she might handle it differently.

- Remember emotions cannot be right or wrong; focus instead on whether the actions we take as a result of our emotions are helpful or unhelpful to the situation in hand.

CHAPTER SEVEN

Attachment: why emotional regulation matters

Your capacity for emotional regulation plays a part in shaping your child's ability to regulate emotions and there are a number of reasons why this is the case, which I'll outline here. As humans, we have evolved over hundreds of thousands of years to be biologically endowed with the capacity to love and respond to our babies in ways that nourish their bodies, brains and minds. But as we already know, various aspects of modern-day life, including the increasing left-hemisphere dominance in the way we live, interfere with what is designed by evolution to be a natural and intuitive process between parent and child. In this chapter I'm going to outline why and how emotional regulation shapes us through a process called attachment, which has a strong impact on how we come to feel about ourselves and how we later relate to people with whom we have close relationships. In the next chapter I will outline the different categories of attachment and how they play out in day-to-day parent–child interactions. But first, let's explore the link between attachment, emotional regulation, the parenting relationship and your child's sense of himself.

ATTACHMENT IS THE MEDIUM THROUGH WHICH WE FORM RELATIONSHIPS

Human babies arrive in the world with highly immature brains and bodies because of which they are vulnerable and highly dependent on a caregiver for their survival. In very simple terms, if a caregiver is not attentive to their needs and is not moved enough by their interactions with the infant, they may neglect to care for the infant, which will result in death. This is not something babies consciously know but it is hardwired into their genes and they are therefore born with an inbuilt sensor system that monitors not just whether a caregiver is in close proximity to them but whether that caregiver is tuned in to and responsive to their needs.

Babies are wired to want to be close to us for many reasons

Proximity matters so enormously that separation from a baby's primary caregiver causes physical stress in their bodies, something they release through intense crying designed to make their caregiver relinquish any notion of doing something other than holding their baby close. We all know how difficult we find this at times but it is a minor discomfort compared with the survival stress that a baby faces in being completely dependent on another human being who may or may not notice and/or care about what she needs. In days gone by, proximity was a natural counterpart to early parenting because we didn't have the apparatus with which to make them independent of us. We wore them in slings because we had to; we all slept huddled together in the days when we weren't so concerned with space, control, independence and comfort; we walked close together, we cuddled for warmth and comfort and all this proximity increased the chances that we noticed and picked up on what our children needed from us.

Children need proximity and emotional attunement to feel physically and emotionally safe because they don't have the brain parts responsible for self-regulation when they are born. When they are distressed, they can't consciously bring their heart rates, breathing, stress hormones and muscle tension to a state of equilibrium or comfort by themselves. They are dependent on you to do this for them. Their needs, and your responses to those needs, are expressed through emotions and because of this your baby arrives with the ability to notice and synchronise with your emotional states from the time she is born.

The process through which this reciprocal emotional relationship becomes established and secure is called attachment. Researchers have established, over many years, that we can have what is called an attachment type that is based on the characteristics of our relationships with our primary caregivers. Please note that a child can demonstrate a different attachment pattern with each parent and that a parent may also have a different attachment pattern with each of his or her children. Attachment is not a trait but rather a relationship pattern that might play out differently between people depending on the individual personalities, temperaments, qualities, insecurities and triggers in each relationship. Some attachment styles may be quite strong and may be observable across multiple relationships. In others, their attachment style may not be so clearly defined. What's more, attachment patterns can also change over the years so none of this implies anything is set in stone but it is enlightening to reflect on your own pattern and what it means about the way you engage with your children all the same.

Attachment is about emotional reciprocity

Attachment is a fundamental human drive that babies are biologically hardwired for and it is through the interactive regulation of emotional states between people that attachment processes play out and relationships are formed.[34] This is really about **emotional reciprocity** or 'matching of emotions' between the primary caregiver and the child such that the caregiver and child come to synchronise their biological systems, for example heart rate, nervous-system activity, stress chemicals and facial expressions. When this occurs, and the caregiver is in a regulated state within their window of tolerance, the baby feels safe and comfortable. Because dysregulated emotion can feel so uncomfortable, the emotional security that effective coregulation provides is invaluable to children. This is not something that you must 'do' but rather a state that arises intuitively from the capacity for empathy and emotional openness enabled by a balanced right hemisphere. It is also a process that nature has invested heavily in so most of us are biologically wired to produce the hormones and chemicals, such as oxytocin, that make bonding with a baby or a partner pleasurable and rewarding.

Please bear in mind that such emotional reciprocity will not occur every moment of every day, that would be impossible and draining, but rather occurs in continuous cycles of connection and disconnection. When there is a reasonable degree of genuine connection overall, and the parent–child relationship is characterised by emotional warmth, consistency and openness, the relationship can tolerate a surprising degree of temporary disconnection. When there is disconnection, perhaps because we are preoccupied and don't notice our children's states, or because we are unable to regulate ourselves in that particular moment, we must become aware of this and seek to repair it through re-establishing connection as soon as we possibly can. This is very difficult when we are generally caught up in lists of things to do, or when we are stressed and tired, when we experience depression or anything that causes an imbalance in our emotional regulation neural circuits.

Why do children need parental coregulation?

Because those parts of the brain that enable us to regulate emotions successfully develop after the first eight months or so of life, and continue to develop over time, children have a very rudimentary ability to regulate and soothe their own emotions independently of a caregiver. Basic mechanisms such as sucking their fingers and averting their gaze when they are becoming overstimulated offer them some semblance of control over their emotions at a physiological level, but because the brain parts associated with emotional regulation are slow to develop, they have a small window of tolerance for most emotions. Cue crying, screaming, fussing, whining, shouting and all those things we sometimes label and experience as 'difficult'.

Very young children are at the mercy of their emotions with no real 'off button' to help them soothe uncomfortable bodily responses that cause stress hormones and other chemicals to course through them. The only 'off button' they have at their disposal is you. Your child wants to feel that she can experience her emotions with you under a veil of unassailable safety and that you will help her to regulate them so that she can restore the internal equilibrium that she needs. Negative emotions feel as distressing to your child as they do to you, except your child has even less of an ability to soothe them than you may have. Can you imagine how vulnerable this makes them feel when they and we are regularly out of control? You might find yourself thinking, 'Nonsense. If they felt that bad, they would show it', but remember children often go into hypoarousal, or even freeze from fear and distress.

Attunement happens when we cultivate stillness and acceptance

Real attunement happens when we slow down, stop talking, look into our children's eyes and regulate our reactions to what we feel around them. Where I may have been too emotionally responsive to my children in the early years, I have come across many parents who disconnect too rapidly from what their children are feeling, rushing in to tell them to 'stop it', or 'It's not that bad' or 'It's okay', all of which encourage premature independence in emotional management and a later tendency towards emotional suppression, denial and avoidance.

There are broad cultural tendencies that play a part in how we handle emotions. We British, for example, often regard emotional expression as self-indulgent, destabilising and even a sign of weakness, instead promoting independence far earlier than is healthy. We interpret a child's need for coregulation as a sign of vulnerability that we anticipate will cause problems in our relationship and inconvenience us. We label children 'clingy' and 'needy', and feel embarrassed when they express emotions in a public way. We worry that others will judge our children badly behaved for expressing their emotions. This preoccupation with shutting them down is not helpful to their well-rounded development. But neither is allowing them to express themselves with no consideration for others; it really is a matter of balance.

Even when they are distressed and feeling awful, holding them and soothing them with our regulated calm presence can feel pleasurable to us and them because soothing releases oxytocin, which calms the body and brain. This type of regulation is an authentic moment-by-moment process that emerges from a sense of stillness and softness in the mind and body, something that is shaped by the functioning of the vagus nerve that I'll describe in a later chapter. It does not require that you read or follow any techniques designed to tell you how to speak to your child or what to do. It requires a mindful paying of attention to what is happening without any judgement or preoccupation with what things mean, what 'should' or 'shouldn't' be happening or even what we need to do next.

A tuned-in parent notices when their child is becoming dysregulated from their eye contact and body language, and may intuitively back away if they sense their child is not responding synchronistically to their interaction, whereas a parent prone to hyperarousal may fail to notice the child's subtle responses and intrusively impose themselves on the child, continuing to talk loudly, force the child to engage, or do something that the child has either to resist or then feels powerless to stop. On the other hand, parents experiencing hypoarousal may fail to pick up on the child's need for soothing entirely, instead responding to the child with a somewhat detached or deadened expression in their eyes (they may be attempting to speak in an animated voice but the eyes can't fake emotions). Children often attempt to self-regulate through avoidance of something challenging and although this is quite normal for them at that stage, some parents don't acknowledge this and instead cajole or push their children into doing what the parent wants, and in doing so deny or ignore their emotional state.

YOUR CHILD'S EARLIEST SENSE OF SELF DEPENDS ON NON-VERBAL FEEDBACK FROM YOU

In the first few weeks and months after they are born, babies cannot differentiate themselves from their primary caregiver. For the sake of convenience and based on the law of averages, I'm going to assume that the primary caregiver is the child's mother. Babies do not know and cannot understand that they are separate entities from their mothers so their emotional reactions and those of their mothers seem fused together. Because babies are primarily reliant on right-hemisphere non-verbal language for the first year and all the way up to the age of two or three, the mother's emotional reactions, borne out in changes at the level of the nervous system, heart rate and micro-facial expressions, provide valuable information to the child as to how to interpret and feel in his environment, including about himself.

Emotions are the lens through which he learns whether he is safe or vulnerable, cared for or neglected, validated or judged, and the only way he can know 'who he is' is from what he perceives are his mother's emotional reactions to his presence. He learns about what he is feeling, and the essential 'rightness' of that emotion from what she mirrors back through her own reactions to him. Her emotional signals provide the child with feedback not only on his own emotional states but also how he is perceived by her, something that shapes how he comes to feel about himself. Because of this, her ability to tune in to and notice his emotions, moment by moment, is a vital part of this process.

Whether or not this is possible will depend on the attachment style of the parent. Emotional regulation plays a significant part in this because it is through the language of emotion that attachment patterns become established and stored as 'internal representations' in your child's brain that essentially become a model for future relationships with others. Essentially, your child internalises a series of images/emotions/thoughts in relation to how lovable he is to you. When your child thinks about you in relation to himself, will he have in mind an image of you with a warm, smiling, kind face or a cool, distant face, or a critical, contemptuous face? Children begin to form beliefs, held out of conscious awareness, about themselves, what they deserve and what they are capable of, from an early age, much of which is shaped by their interpretation of their experiences. Examples of such beliefs may be:

Negative	Positive
– I'm not lovable enough	– I am lovable as I am
– I'm not good enough	– I am good enough
– I must please others to be accepted	– I can be myself; do what feels right to me
– I must get things right all the time	– I can make mistakes
– I can't rely on people to be consistent	– I can trust people to be there for me

Children develop a positive and stable sense of themselves, what we call a self-concept, based on the reactions they perceive others have to them, and no one more so than the person they rely on for emotional regulation. Children need to feel validated and 'seen' on the inside to grow into authentic beings who don't just exist through the eyes of others. When this doesn't occur, as we established in the previous chapter, they may either become too self-sufficient and independent as a form of self-protection or they become too reliant on the approval and feedback of others to shape who they are. They may even become overly driven by extraneous sources of validation and satisfaction such as achievement, status, control or pleasure.

Many parents think praising and attending to their children constantly will create a strong sense of self-esteem but this only encourages them to objectify themselves. Rather than tuning in to how they feel about things they are doing, for example, whether they feel curiosity, motivation, anticipation or self-satisfaction, they begin to want to do things because of how it feels to be praised. Children don't need praise and rewards to feel this sense of validation; rather what they need is to be able to feel and express their authentic emotions in a way that is noticed and reciprocally accepted and regulated, through the process of parent–child connection. I'm not saying don't praise them at all; it may feel perfectly natural and encouraging at times. What I am saying is don't assume that praise is an indicator of connection.

HOW WE RESPOND TO OUR CHILDREN'S EMOTIONS SHAPES THEIR MENTAL MODELS ABOUT EMOTION

Successful coregulation occurs when you empathically observe your child's distress without becoming either overly distressed or disconnected. Parental coregulation of emotion works on two levels:

- It builds those brain regions that later help them to regulate emotion.
- It implicitly gives children messages about emotions, and their ability to handle them, that they internalise in their implicit memory system.

These internalised messages gradually build up into a set of beliefs or a 'mental model', which is an amalgam of all the information we have stored and processed in the brain about a particular object or topic. Children develop mental models about themselves, about emotions, and also about how people will respond to them in times of need, something that shapes how they deal with their emotions in later life. In many cases, these models become persistent characteristics that we may not see as having been influenced by early experiences but rather as being 'who' we are. These models shape us outside of conscious awareness and can often explain things that perplex us about ourselves.

When you are regulated and compassionate in the face of your child's distress, when your eye contact is gentle and your touch is soothing, her stress levels drop and her heart rate and breathing return to their baseline. When this occurs regularly, the child develops a sense of confidence in being able to handle uncomfortable emotions because she learns that no matter what she is feeling now, she can always feel okay again later. This child will develop a set of beliefs about emotions that might be comprised of these:

a.   emotions may feel uncomfortable but they are normal and tolerable in general – they don't need to be afraid of them because they can calm their emotions down.

b.   their emotions, and therefore they themselves, are valid and okay.

c.   they can express their feelings with a sense of trust that others will understand them and help them to manage uncomfortable states.

As a result of this, they don't need to act out by vociferously vocalising emotions (hyperarousal – fight) or to suppress them by avoiding vulnerability (hyperarousal – flight) or by dissociating from them (hypoarousal – freeze).

On the other hand, when we respond in a dysregulated way to their expressions of emotion, especially if we are critical or unsympathetic, what they learn is that their emotions are wrong or dangerous and will either overwhelm them or leave them feeling alone in the experience of having them. This can lead to a tendency towards hypoarousal or active suppression of emotions which is a state of hyperarousal. Holding beliefs that emotions are unacceptable or that expressing them will lead to negative consequences has been associated with several problems such as depression, eating disorders, chronic fatigue syndrome and irritable bowel syndrome. For example, if you struggle to regulate the emotion of sadness, and tend to either make light of things – such as using humour as a distraction, or suppression and avoidance to detach yourself from painful emotions – or even use comfort mechanisms such as food and pleasure as distractions, your child internalises a sense that sadness is unbearable and to be avoided. Your child may later have a tendency to avoid feeling those emotions, perhaps taking pride in putting on a brave face, being strong or independent, and seeing their own vulnerability and emotion as a sign of weakness. They may also find themselves feeling emotionally numb, only capable of experiencing mild emotions on the pleasure spectrum. They may learn to exert self-control but not to connect with people in a meaningful way. Some may even attach greater importance to helping others and putting their needs first, whilst learning to ignore and suppress their own.

You don't have to respond to your children's emotions all the time

Although tuning in and helping them regulate their emotions is critical to them developing a healthy sense of themselves as individual beings in your eyes, and therefore increasingly their own, this does not mean that you validate every emotional reaction they have to something. This must not be some parrot-like trotting-out of phrases such as 'I know you feel sad' every time they demonstrate an emotion; this won't be authentic and comes from the left brain rather than the right. When you are naturally tuned in with them, empathic phrases such as 'That must be so annoying right now' emerge naturally, are spoken in a gentle, casual manner and sound genuine. Most of the time, you don't even need to say much to resonate with what they are feeling. When you feel it inside, they sense it and they respond by calming down. Sometimes you just need to let them be.

Emotional resilience is about helping them tolerate uncomfortable states

When you help them to tolerate what seem to be intolerable feelings, they develop emotional resilience. It is not through avoiding difficult emotions or protecting them from discomfort that we build resilience in our children; it is when we are able to be in a regulated state with them whilst they experience dysregulation that we empower them to be adaptable in later life. This means your child will be able to move in and out of different situations with greater ease, staying within their windows of tolerance and employing appropriate emotional and behavioural responses to adjust to each new experience. Each time you calm a child who is dysregulated you are building the neural pathways for successful self-regulation in later life.

Of course, your job is to protect them from very destabilising emotional experiences as much as possible but many of the experiences that parents worry about these days are really not things to be shied away from at all. Our children do not have to be protected from the emotional discomfort that stems from social rejection, from ad hoc criticism, from missing out on things their peers may experience, or from failure. It is possible for them to like and accept themselves regardless of those things. On the contrary, I believe we are sabotaging their capacity for resilience by being overprotective and they are growing up with a sense of entitlement and expectation of things that simply cannot be guaranteed until we live in a perfect world full of perfect human beings. When these internal rules and expectations they have generated are violated, it creates unnecessary anxiety, stress and unhappiness.

There are also ways to promote mental resilience that involve conscious verbal strategies such as reframing thoughts, challenging negative interpretations and other such methods that I will outline in a later chapter. These are techniques that you can role model and actively teach your children, though I would advise you to resist the temptation to rush into these rationalising techniques too quickly. When it comes to regulating emotion, it is best to get a balance between feeling our emotions, soothing them and thinking about them constructively in order to solve the problems that may have led to them.

STAYING CONNECTED YET DISTINCT FROM YOUR CHILDREN

The type of balanced emotional regulation that allows us to return easily to our 'window of tolerance' requires integration between various brain parts in the right and left hemispheres, and also integration between the body and the brain so that information can travel along all the relevant routes and recruit all the necessary resources along the way to reach a successful outcome. When we have strong and healthy connections between these regions without too little or too much neural activity or connectivity, we are able to feel our feelings with authenticity but to also differentiate our emotions and experiences from others around us, something that is critical to parenting.

Differentiation means being able to separate the different strands of emotion in oneself, for example, knowing you are feeling frustrated and sad at the same time. It also means you can recognise and feel another person's emotional state whilst distinguishing your own emotional response from what that person might be feeling. Parents who struggle to differentiate their feelings or sense of selves from those of their children tend to impose their own emotional states on their children or become easily triggered by their children's emotions. Highly anxious parents, for example, often struggle to empathise and resonate with their children's emotions without becoming dysregulated by them. Rather than quietly tuning in to your child's emotion from his perspective, you may become led by your own emotional reaction to what you have assumed your child is feeling, based on what you might feel in such a situation. Or you become caught up in some perceived injustice that you believe your child has fallen prey to. If you can successfully differentiate yourself from your children, you can feel connected and moved by your child's emotions but can remain regulated and independent at the same time. Parenting feels so personal at times that it becomes difficult to observe without rushing in to judge or react.

Exercise: Accepting negative emotions in your child

Please do this exercise after you have practised the exercise in the previous chapter a few times. Start by finding a quiet space where you are unlikely to get distracted or disturbed. Sit with your body in an upright and comfortable position, ideally with your arms and legs uncrossed. You might like to close your eyes so you can really focus on the interior of your body. Take a few moments to tune in to yourself and your inner sensations as you breathe in and out, whatever they might be.

1. **Recall** – Bring to mind a situation where your child was dysregulated with a negative emotion that you find hard to tolerate. Let yourself fully imagine it without trying to sanitise it in any way.

2. **Experience** – Now notice how you feel when you imagine your child out of control. Let the emotions you felt during this episode

unfold without trying to do anything about it at first – just notice the four markers of emotion in turn.

- Heart rate – does it go up or down?
- Breathing – can you notice any changes to your breath?
- Muscle tension – pay attention to the muscles around your heart, chest, gut, face and throat.
- Chemicals/sensations – what do you feel and where do you feel it?

This may be very hard for some of you but do persist. Instead of tightening and tensing up, or judging and analysing the emotion, just let it be for a few moments. Try not to react to it.

3. **Soothe and calm** – Now, once you've tolerated the emotion for a few moments, start to calm and soothe it using your breath. Keep the image of your child out of control in your mind at the same time if you can. Take a deep inhale up into the space behind the middle of your forehead and breathe out into your body; let your body relax as you exhale. As you breathe in, allow your belly and upper back to expand with the breath and exhale deeply in whatever way feels comfortable to you. Try not to tighten your body because you're trying too hard. Repeat this up to five times or more.

4. **Tune in to your heart** – Now continue to inhale up into the middle of your forehead but exhale out into your heart and the space around it. Take your full attention to the area around your heart and the sensations there. Use your breath to soften and relax that area. Imagine your exhale is a form of kindness and compassion for your body and breathe this sense of gentleness into all those tense muscles around your chest. Slow down your exhale so it is longer than your inhale. Repeat this with your belly.

5. **Notice what's changed** – Focus your attention on the parts of your body that need to be soothed and calmed the most. Soften

your face and uncrease all the tiny muscles there. Exhale into the whole of your upper body, so that it feels as though your breath is almost travelling down from the space behind your forehead into your heart and gut. Notice any changes that have happened. Keeping an image of your dysregulated child in your mind, try to turn the outer corners of your mouth up into a tiny fraction of a smile.

6. **Engage your mind** – Once the sensations of the emotion in the body have lessened, try telling yourself something calming as you exhale. You might like to say, 'It's okay, it's just a child experiencing an emotion – I can let this go.' Or something similar that allows you to put it into perspective. Remind yourself that your child needs your support not your judgement.

7. **Repeat the exercise** – Go through the exercise again and repeat until you feel reasonably calm when you bring to mind your trigger situation. It helps to use this exercise with trigger situations involving your children, as repeated practice will teach you to react less in real life.

Reflection: What are your beliefs about emotion?

- Do you believe that expressing a range of emotions is normal and healthy?
- Do you believe that emotions are manageable or do you fear becoming overwhelmed by them?
- Do you believe that people can change their emotional responses through self-soothing or mental effort?
- Are you able to feel vulnerable without becoming overly distressed?
- Do you sometimes feel the only way to handle strong emotions is to suppress or deny them?
- Are you able to accept uncomfortable or negative emotions in yourself and your children?
- Do you find yourself reacting in a dysregulated way to certain displays of emotion from your children, for example, crying, anger?

- Do you look for logical reasons to explain your or your children's emotional reactions to yourself or others?
- Do you judge or criticise yourself or your children for having strong or irrational emotional reactions to things?
- Do you see emotional expression as a sign of weakness?
- Do you think others will reject you if you show vulnerability or express negative emotions?
- Do you believe you should cope with your emotions silently or independently?
- Are you able to savour positive emotions such as gratitude and awe?
- Do you feel emotions at a bodily level – are you moved by them or do you tend to think about them and analyse them? Is this the case with just some emotions or all?

Chapter Seven: Key points

- Attachment is the process through which emotional regulation and attunement between parents and children shapes a child's self-concept and how they engage with those with whom they have close relationships.
- Babies are genetically hardwired to seek proximity and responsiveness from their primary caregiver. A lack of proximity or responsiveness causes survival stress, which facilitates the release of stress chemicals that feel deeply uncomfortable.
- Attachment is primarily about reciprocity – the matching of emotions that occurs between a parent and child when they are in tune with each other. Attunement happens when we slow down, feel connected with our emotions, and match our emotional responses with what the child most needs from us.
- Children need to have their emotions, ideas and actions validated through observing your responses to them in order to develop a coherent self-concept. This doesn't occur through praise and adoration but rather from feeling understood and accepted for who they are.
- Emotional resilience is built by helping your child to tolerate

and regulate emotions. This means you mustn't feel the need
to protect them from feeling uncomfortable or distressed as a
result of rejection, criticism, failure and general disappointment.
Instead, focus on helping them to notice, soothe and reflect on their
emotions.

- We must be able to differentiate our emotions from those of our
children. Differentiation means being able to separate the various
strands of emotion in ourselves and others, and being able to
distinguish our own feelings from those of others to whom we feel
close.

CHAPTER EIGHT

Your attachment style frames the way you parent

Now that we understand how emotional regulation shapes our children, let's look at the way in which this affects their attachment style with you and yours with them. Attachment can broadly be characterised by whether it is based on security or insecurity, with insecurity subdividing into two further categories. Based on this, attachment can be defined as **secure, insecure ambivalent, insecure avoidant and disorganised**. Attachment categories have an influence on the way we conduct ourselves in close relationships but we also know that regardless of their early experiences adults can move from an insecure to a secure attachment style, either from experiencing other consistent, loving relationships in their childhood or later in life, or through psychotherapy or deep self-reflection over time. Do also bear in mind that we may have different attachment styles with different people, for example with our parents and romantic partners, and children may have a different attachment style with each of their parents. The child–parent attachment style has an impact on how well-adjusted and happy children are in general.

THERE ARE FOUR MAIN ATTACHMENT STYLES

Secure attachment – When parents are tuned in, emotionally available and responsive to the emotional and mental states of their children more often than not their children develop what is called a 'secure attachment style'. People with a secure attachment style tend to trust people, see the world as a safe place, and have confidence in their ability to handle their emotions, either on their own or with help from people they are close to. They are not afraid of vulnerability or intimate connection and don't worry too much about rejection or disapproval. This is not to say they don't feel upset when these things happen but they don't have an expectation of it before there is evidence that it will happen, and when it does it does not strongly knock their sense of themselves as individuals.

Securely attached people have a wider tolerance for emotions and are usually able to regulate their emotional states with resilience and flexibility, but also to have a reasonable degree of empathy for others around them. They don't tend to become overly anxious and are open to new experiences and opportunities (though this is also influenced by many other things including other personality traits, not just attachment styles). A securely attached child sees herself as lovable, worthy of attention (but not needing too much of it) and generally capable of having good relationships with others. Secure attachment forms the base from which your child learns to approach others for emotional support when she is distressed or uncomfortable, and she will have an expectation of being able to resolve her problems and upsets when they arise. Securely attached people share their emotional states with others when this is appropriate, tend to find their personal relationships satisfying and express greater contentment with life in general.

Insecure ambivalent – Children who develop this attachment style usually have parents who are emotionally inconsistent. It arises from a parenting style that is at times unavailable and preoccupied, and at other times, emotionally misattuned and intrusive. Parents with this attachment style tend to become easily triggered by emotions in themselves or their children and struggle to regulate these states. They frequently interfere with what their children may be doing. They respond to their children in line with what they need to do to relieve their own emotions rather than what the child needs in order to feel comforted and safe. Such parents may think they are being loving and responsive but their emotional responses are often not tuned in with the child's inner state or needs, in which case they can be intrusive and overpowering, and can heighten the child's dysregulated state. When people with such an attachment style feel overwhelmed, they may have outbursts of emotion that can be inconsistent or chaotic.

Examples of this style of parenting might be interrupting the child whilst he or she is speaking, or speaking over him; continuing to talk to a child about a topic that the child clearly cannot tolerate speaking about because you need to let your thoughts and feelings out; talking to your children in an overly animated voice when they are feeling quiet, sad or low; grabbing, kissing and holding your child when she is preoccupied with something else because you feel an urge to connect; interrupting or finishing what she is doing at the first sign that she is struggling with a task; or convincing the child to do activities because you think they need or want them without really tuning in to whether they enjoy those things or not. In other words, parents who don't really 'see' the child as a separate person with different thoughts, feelings and needs but whose emotions, expectations and responses to their children stem from their own inner emotional lives.

Parents with an ambivalent attachment style can tend towards 'helicopter parenting' and other protective styles of parenting where parents are very involved in their children's lives, schooling, friendships and experiences. Some of these parents tend to talk extensively about what their children are doing, their activities, achievements and successes, taking a lot of pride or interest in them and becoming too personally invested in their children's experiences. In extreme cases, the parent devotes his or her life to the child's comfort and advancement, creating unseen volumes of pressure on the child and scuppering opportunities for relaxed connection.

Parents with this style struggle with differentiation and can't fully separate their own emotional state from that of their children, which in turn creates children who become overly vigilant about the emotional states and reactions of the parents. Both parents and children with this attachment style tend to take things personally, perhaps overreacting to what they perceive to be unfair treatment of themselves, and interpreting things in a personal light – 'it's about me or my child' – where other possible explanations might better fit the facts. People with this attachment style tend not to view themselves as wholly lovable and safe, therefore remain sensitive to signs of criticism, rejection, unfairness and abandonment.

Children often internalise the insecurities and anxieties of ambivalent parents and organise their self-concepts around pleasing or calming them. For these children, there is little confidence that their emotional needs will be consistently met by their parents and in some cases, the child becomes preoccupied with attracting and maintaining the attention of attachment figures so that they know someone is available to help them regulate themselves should they need it. They need and demand attention and coregulation but may show conflicting patterns of emotions such as anger and anxiety, and are not always easy to soothe. They desperately want connection with the parent but also refuse to be soothed by it and may even push it away and reject it, making it hard for parents to figure out what the child really needs. Some parents deal with this rejection by taking it personally and distancing themselves from the child emotionally in the moment, creating even more confusion in the relationship. Others deal with it by trying too hard to please and be loved by the child, which might fuel insecurity in the child.

Insecure avoidant – This attachment pattern is characterised by emotional unavailability and dismissiveness. Parents with this style tend not to notice and respond with warmth and acceptance to their own emotions or the emotions of their children, especially when the child is dysregulated. There is a tendency and a need to maintain autonomy and emotional distance from people, taking pride in being independent and sometimes seeing people who get too close as needy, imposing and an inconvenience. People with this style are likely to have had parents who were emotionally unavailable or rejecting of their need for comfort when they were distressed. As a result, they have learnt to regulate emotions in an autonomous way, not expecting to share their feelings with others easily and not really knowing how to open up. They tend to have a logical, cool and distant way of looking at things, may be controlling and have a preference for drier forms of wit.

In terms of emotional regulation, parents with this style have a tendency towards either actively avoiding, suppressing or denying emotions, possibly remaining in a mild state of hyperarousal on a regular basis. They may struggle to relate to all this information about emotions and psychological needs because they really don't receive and process emotional signals from within themselves or from others around them and therefore cannot fathom what the fuss is all about. Because they tend to want to suppress certain emotions and find them hard to tolerate, having a child who experiences intense emotional states, as they all do when they are very young, is perceived as an imposition and even an intrusion on their need for space and autonomy as adults. Parents who have these insecure attachment styles may frequently feel stressed by being around their children.

Parents and children with this attachment style tend not to divulge their emotions easily, especially emotions on the vulnerability spectrum. Their eyes and faces don't reveal much of an inner emotional life moment by moment and in extreme cases can seem a little flat and devoid of warmth, making it harder to 'read' them. At a very deep level that many never acknowledge, they don't trust themselves or others to handle their feelings sensitively, and they protect themselves from this by shutting off emotions and going through life with a level of emotional detachment. Please note such individuals may be fully capable of experiencing and expressing pleasure, excitement, drive, humour and emotions on the anger spectrum, but not sadness, vulnerability and real emotional intimacy.

Parents who are avoidant encourage their children to become independent, toughen up and manage their emotions by themselves at an early age. There is a tendency to want to distance themselves from their children, especially when they are in the midst of intense emotions, and to achieve this they may switch off internally by tuning out, or they might physically distance themselves from the child temporarily or in some cases more permanently. Where parents with the ambivalent style might worry too much and overreact to their children's distress, parents with the avoidant style don't worry enough about how their children feel. They struggle to understand when they need to be appropriately involved in their children's lives and may prefer to leave the child to sort their own problems out, often under the guise of 'toughening them up' and not overprotecting them. This is not because they are uncaring or unkind but because they are generally less aware of their own and others' emotional lives.

Parents with the avoidant style are often out of synch with their children. In everyday life, this might manifest itself in simple forms of misattunement, for example, when a joyful and excited child runs up to his mother to say, 'Mummy, look what I found', and before he can even finish the sentence, his mother, not looking into his eyes, says dismissively, 'I'm in the middle of something. Go and play.' Or, a parent might suddenly leave a room with her child in it without understanding that her child will feel distressed by her absence. Of course, we all have moments like this but these attachment patterns reflect frequent instances of behaviour between parent and child rather than occasional ones.

The avoidant pattern reveals itself in whether the parent notices or understands that the child might perceive the parent's behaviour as a rejection and feel a sense of shame or anxiety. Such parents often have practical or negative explanations for the emotional behaviour of their children, such as a child being hungry or thirsty or 'silly', and they may tend towards rigid routines and discipline rather than flexibly adapting what they do based on the situation and emotional needs of the child in that moment. When their children are distressed, parents with an avoidant style may find themselves feeling quite impatient and frustrated with their emotional outbursts, saying things like, 'Come on, you're fine. Sort it out' or, 'You're eight years old now, you can cope with this.'

Disorganised – This category is perhaps the most troubling and the least common. It tends to be the style exhibited by the children who have suffered some form of abuse, severe trauma or even unresolved childhood loss. It has elements of the other two insecure styles but with a lack of any coherent pattern and strategy for dealing with emotional and relational needs. People with a disorganised attachment tend to swing between both hyper- and hypoarousal, with states of complete blankness at times akin to a total shutdown. This pattern is also associated with parental neglect; extreme and frequent outbursts of anger or emotions that frighten, shame and humiliate the child; and substance abuse such as alcoholism and drug addiction. Being on the receiving end of such behaviour is traumatising for the child because the child looks to the parent for security against a threat, and when the parent is the source of threat it leaves the child feeling confused, defenceless, alone and vulnerable. Please do remember that this is about regular mistreatment rather than the odd episode of anger.

Earned security – We are not destined to stick with our childhood attachment pattern for life but are capable of earning a secure status as adults through experiencing loving relationships with other people along the way. We can heal ourselves through therapy and increasingly, I believe, through learning how to be compassionate towards ourselves and others. As Daniel Siegel points out[35], 'although attachment patterns tend to be transmitted across generations, the best predictor of a child's attachment status is not what happened to his parents when they were children, but how his parents made sense of their childhood experiences.' This means being open, aware and reflective of our early experiences and the impact this has had on us can help us to move away from an insecure style towards a secure one.

SHAME, EMOTIONAL REGULATION AND ATTACHMENT

It is often challenging to recognise and validate the primary, driving emotion behind a child's distress from the child's frame of reference rather than your own. It is only when we are emotionally regulated that we can tune in to the full spectrum of a child's emotional response without needing to react to it immediately. When you are able to stay in your window of tolerance, because the different brain regions that allow you to pay attention in the moment and take in new information are still connected, you can think about possible reasons and causes for your child's emotions that are not unduly negative or personal. When you are able to be open, calm but connected at the same time, your child feels understood and will therefore allow himself to be soothed. When there is a misattunement, which frequently occurs when we don't validate or correctly understand our children's emotions – for example when we jump to conclusions about their behaviour and label or judge them – they will feel ashamed and, perhaps after an initial angry defence, will shut down (hypoarousal).

Shame, which Allan Schore[36] describes as a feeling of 'inner collapse' involving a rapid transition from a positive state to a negative one, feels deeply uncomfortable for children and they interpret it as a sign of rejection. To them it can mean they are bad, wrong, unlovable, not good enough. It can quite easily happen from misattunement and starts to occur frequently once toddlers begin to push boundaries and need to be taught appropriate social rules through correction. The example I gave above in the section on the avoidant parenting style describes a shame-inducing scenario when a child comes to a parent full of joy and she rejects his bid for connection in a cool or dismissive way. Children also feel shame when they are harshly told off and corrected, especially if this persists for a length of time. If this happens more often than positive interactions, or is handled insensitively on the part of the parent, it is painful enough to cause a child to turn off his feelings, his need for emotional coregulation and also his need for closeness.

The feeling of rejection or shame can also lead to rebound ego-defensive anger or defiance because this feels safer than experiencing the vulnerability and pain that feelings of being misunderstood or unworthy can involve. Children vary in their sensitivity to these emotional states.

Some can feel ashamed by something as simple as a parent angrily telling them off; others may tolerate a higher level of misattunement and criticism before they begin to feel ashamed or defensive. One sign that your child is feeling a sense of shame or misattunement is when they stop making eye contact with you and can't look into your eyes as you're speaking. If this is the case, it is a sign that you need to soften your approach or even stop talking and let the child return to his or her window of tolerance. If you are telling your children off angrily and he begins to cry, that is also a sign that you have gone too far and it has evoked a feeling of shame, sadness or rejection. If he is arguing back, refusing to listen and being defensive, it means he is feeling disconnected and misunderstood and he is defending himself from a threat, which in this situation is being misunderstood by you.

As a parent, you need to help your children to regulate these feelings because they are really hard to handle and feel unpleasant. If you yourself have an insecure parenting style, this might be challenging for you but you must remain open and aware that you might be overreacting or making negative assumptions about the causes of your child's behaviour that will lead to excessive shame and rupture in the connection between you. This could lead to the child shutting down, in which case he won't really be listening to you, or him becoming belligerent, which will exacerbate the emotional dysregulation between you. But in either case, all potential for a good outcome is lost because once the child is out of his window of tolerance his ability to take in information with an open mind is compromised.

Here is an example of a conversation between a parent and child that has created a sense of mild shame and misattunement:

Parent (in a vexed tone of voice): 'Your teacher mentioned you haven't been doing your homework on time lately. You know you'll get into trouble if you don't do your homework on time. Why aren't you taking it seriously? Have you even done your homework today?' Here the parent is asking questions not out of open-minded curiosity but out of judgement.

There is an expectation that the child 'should' have done the homework by now and it reveals itself in the tone of voice used.

Child (in an irritated, defiant, slightly raised tone of voice): 'I have been doing it! I only forgot once. I'll do it later.' The child feels a bit judged and controlled, and thinks the question is unfair.

Parent (a bit annoyed now, intruding on the child's need for space and time to finish what he is doing): 'You know you won't feel like doing it later. I'm sick of always having to tell you to get on with it. Why can't you just take responsibility for yourself?'

Child: 'I will. But later. You don't need to keep telling me to do it.' What the child is really trying to say is 'Please trust me and respect my need to finish what I'm doing right now.'

Parent (knowing he should drop it or take a kinder tone but unable to control his frustration because he has tipped out of his window of tolerance, starting to shout a bit now): 'But you haven't been doing it! You're too interested in other stuff! You'll only forget later and then I'll have to keep reminding you. You'll get into trouble at school again for not doing it.'

Child (feeling annoyed at the unfairness of being reminded about getting into trouble when he was actually intending to do the homework later): 'Fine. I'll do it NOW!! You always force me to do things when I know I need to do them anyway.'

Parent (now calming down and returning to his window of tolerance): 'I'm not saying you have to do it right this minute. Just that you mustn't forget.'

Child (still angry and feeling misunderstood and not making eye contact): 'No, I'll do it right now. You said do it now so I *will* do it now!' Although the child has agreed to do it, this is not a positive outcome because the child feels coerced, criticised or annoyed and the connection has been ruptured. This, especially if it has occurred before, may lead to an unhelpful association in the child's mind between the emotions of shame and anger and his homework.

In this example, the parent has jumped to conclusions and his worries, perhaps of the child not doing well at school, or that he will have to get involved in reminding the child again later, have biased his approach. A more fruitful approach might have been to maintain a gentle, relaxed tone and ask an open question, such as, 'Your teacher mentioned you haven't been handing in your homework on time lately. What's led to this?' If the child responds with defiance, the approach must be softened even more. The key is to let the child describe what has happened from his perspective. This neutralises the threat of judgement and allows the child to open up and be more honest. Then comes the problem-solving part. To help the child learn how to take responsibility for himself the parent needs to avoid telling him what to do or coercing him into doing the homework now. But instead, to ask when he is planning to do it. The parent might also ask if a gentle reminder will be helpful or whether he is confident he can remember to finish it himself. The focus needs to be on finding a solution together that addresses the problem in a way that is acceptable to both parties and doesn't feel belittling or coercive to the child. Most of us, even children, have an inbuilt hardwired need for autonomy and fairness.

To avoid too intense a feeling of shame, fear or a defensive response, when disciplining children lower the volume of your speech and speak to them slowly with gentleness in your tone of voice and your facial expressions, but clearly and directly convey a set of behavioural expectations and consequences where this is appropriate. We don't need to judge or label our children (this stems from our own anger response) but simply describe the behaviour we don't approve of and tell them what we do or don't expect them to do. The more connected yet in control you are, the more your child will listen and feel relieved.

It is important to note that you must not swing the other way and try to avoid interactions with your child that might lead to shame. If you try too hard to protect them from it, they won't develop resilience and may lack an understanding of boundaries, consequences and the implicit social rules that breed collaboration and trust between people. They might develop a sense of entitlement and their capacity for moral reasoning could be compromised. Children must not be protected from feeling bad but we do need to be aware of how they are feeling when we are correcting them, and maintain a level of engagement that allows them to tolerate their own feelings. We can push them to the edge of their window of tolerance without pushing them out of it entirely.

Let your children tell the 'story' of the ruptures in your relationship

When ruptures inevitably do occur in the relationship, especially over discipline and similar issues, even if it is for a few minutes, where possible we need to repair the disconnection by reconnecting non-verbally and also by talking about what happened and how they felt at the time. Do note that your need to repair the rupture may feel intrusive to your child and must be timed so that both of you want it and can handle how it makes you feel. To do this, first focus on reconnecting by showing empathy and seeking closeness. Don't be afraid to show your child you understand how angry or upset she must have felt when you lost your sense of perspective or your temper. Apologise to your child with a sense of humility and kindness. Talking about these things helps them to process their emotions and they are much less likely to linger in your child's implicit memory system, affecting their sense of themselves and their relationships at a deeper level.

Let your children tell the stories of how upset they felt at your behaviour as often as they need to, remembering to keep your own feelings of shame or guilt regulated so the focus remains firmly on how your child feels. My two children occasionally narrate stories of the times when I've lost my temper or upset them. Because the stories are quite old, the emotional intensity has now subsided and it's much easier for me to listen to them with a real sense of openness.

But even when the stories were more recent, tough as it was sometimes, I acknowledged to them how hurt they must have felt when those things happened, how distressing and confusing it must have been for them when the person they look to for safety and comfort is also the person upsetting them. And as I heard and accepted what they were saying, and I reassured them that it was not their fault and they didn't deserve that, I saw how validating and healing it felt for them. But I also explained to them that no matter how hard I try, and how much I love them, I am a human being and that means I will sometimes hurt their feelings and get things wrong. And then they'd said, 'That's okay, Mummy' and I'd felt a deep sense of gratitude towards them because I'd shown them vulnerability, and in return, they had shown me strength.

CHILDREN NEED YOUR HELP IN UNDERSTANDING THEIR SECONDARY EMOTIONS

It took me some time to work out that most of the demonstrations of anger from my son when he was between four and seven years old were really a secondary reaction to him feeling vulnerable (sad, rejected, ashamed) and not knowing how to regulate those huge and intense feelings. Children easily become defensive and angry when they feel judged or criticised, especially if they perceive the criticism as unjust. Some parents struggle to accept displays of anger from their children, seeing it as a sign of defiance or disrespect, and taking it personally. Once I understood this and learnt not to respond to the anger but instead showed kindness towards his underlying feelings of sadness or shame, his anger would rapidly dissipate and he would start to cry, which was a healthier reaction for him to have in that situation. I found the sadness so much easier to soothe than the anger and this fits with what we know about the expression of anger – that it evokes more anger.

Over time I taught my son that he has a tendency to ignore his sadness and vent his anger, and he has now become better able to reveal his vulnerability to me without lashing out, becoming defiant or any of those difficult situations that parents dread. The reason he is learning to handle these emotions without converting them to anger is because he increasingly trusts me to coregulate them with him in a way that is attuned and balanced without defensiveness on my part, or any pressure for him to find a solution, to toughen up or even to cover up his sadness because he 'should' be able to handle it better.

TO BE ATTUNED AND EMPATHETIC, YOU NEED TO CARE FOR YOURSELF

Simon Baron-Cohen, in his book on empathy, shares the notion that children have an internal pot of gold that gets filled up when parents nurture and engage positively with their children, bestowing upon them the gift of resilience and security in themselves[37]. The pot starts to fill up with gold each time we experience real connection and are shown warmth, understanding and acceptance for ourselves just as we are. Gradually the pot starts to brim over and we are able to share some of our gold with others who need a boost. When we are frequently criticised, judged harshly, ignored or frightened by those whom we are close to, including ourselves, our pot of gold becomes depleted. It is a struggle to feel compassion, kindness and generosity of spirit towards others, even our children, when we are not able to extend those feelings towards ourselves. Our ability to show ourselves and our children these positive feelings rests on how well it was done for us when we were dependent on others, like our parents and whether, as a result, we have internalised an image of ourselves as lovable and worthy of compassion and kindness. If your pot of gold is lying half empty, unattended and forgotten, no matter how hard you try, parenting will often feel demanding and stressful because you have no inner resources to draw upon.

Please use this to think about how often you criticise and judge yourself, put pressure on yourself, or simply fail to notice what you need in order to feel okay. Give yourself some time to relax and feel cared for. You don't need to rely so much on others to do this for you, and you certainly don't need to rely on your children to fill up your pot of gold, though having loving relationships contributes hugely to it. All my breathing exercises are designed to generate feelings of acceptance and calm that can top you up and replenish your pot of gold. Use them, especially those on compassion, to learn to extend good wishes, kindness and acceptance towards yourself. Start to let go of making rigid demands of yourself as a person and a parent. You'll see how much easier parenting becomes when you practice self-compassion because these states of 'goodness' are not just nice to have, they fundamentally alter how we regulate emotions, our wellbeing, our physical health and our relationships.

Exercise: Generating self-compassion

This is a topic in itself so I'm presenting a brief summary of some of the thoughts, beliefs and feelings associated with self-compassion. My version of compassion is based on the evolutionary perspective, put forward by Paul Gilbert, that we all find ourselves struggling to do the best we can with the brains we have at this point in time. Our brains are shaped by our genetics and our early experiences, none of which is of our making. All human beings will suffer sadness, loss, disappointment, anger and a number of difficult emotions.

Here are some examples of compassionate thoughts:

* I'm having a tough time right now. That's okay because difficulties are a part of the cycle of life. I'm doing the best I can.
* Just like every other human being, I deserve love and kindness, not criticism and judgement.
* I accept myself just as I am – I'm just a complex, fallible human being like everyone else. It's okay for me to get things wrong.
* I'm a living, breathing, multifaceted person – not an object to be judged.

Breathing exercise:

* Take a few moments to tune in with yourself and your breath.
* As you breathe in, take your attention to the space above your eyes (left and right) and as you exhale, turn all your attention towards you face and heart.
* Lengthen your exhale and imagine your breath as a golden light that spreads around your heart. It might help for you to put your hand on

the area of your heart. If you struggle with visualisation that's okay.

- As you exhale, say to yourself quietly and slowly in your mind, 'It's okay. I'm doing the best I can', or 'May I be loved as I am', or 'May I be accepted as I am'.

- Try to really extend a sense of calm, kind, warm acceptance towards yourself. This might feel very difficult if you are not used to showing yourself kindness. You might even cut off from feeling vulnerable at first. Many of my clients have cried the first time they did something like this. Some struggled to stay present because they wanted to go into avoidance mode – their safety net against vulnerability. That's okay. Please try to persist if it doesn't feel too uncomfortable. It will reap huge rewards.

If you would like to explore self-compassion in more detail, Kristin Neff has some wonderful resources you can use.

Chapter Eight: Key points

- There are four main categories of attachment, each of which involves a different way of managing emotions and relationships. Attachment can be classified as secure, insecure ambivalent, insecure avoidant and disorganised. It is possible to have different attachment styles with different people and attachment styles can alter over the course of one's life.

- An attachment pattern is not destined to define you for life as it is possible to change your attachment style and earn a secure attachment status through self-reflection and experiencing stable loving relationships.

- Misattunement, though it happens naturally in all relationships, can cause children to feel a sense of shame that results in a feeling of inner collapse and a move towards shutting down in hypoarousal.

- Pay attention to how your child feels and responds when you need to discipline or correct her, and if she is no longer making eye contact, or has started to cry, or is becoming defensive and angry, take a step back, regulate yourself and soften your approach. But don't be so afraid of inadvertently shaming your child that you shy away from interactions that she might find hard to tolerate.

- Children become resilient when we are able to gently push them to the edges of their windows of tolerance and help soothe them back in when they tip out of it, but you need to be careful not to become a frequent source of negative emotions in your child.
- It is healthy to sometimes experience small ruptures in your relationship with your child through misattunement. Once you and your child are back in your windows of tolerance, try to empathise, open up, discuss the rupture and repair it.
- You are able to nurture and extend positive feelings such as compassion, acceptance, warmth and empathy towards others when your internal pot of gold is topped up. When we experience criticism, have unreasonable demands made upon us, or have our feelings and needs ignored, even by ourselves, our internal pots of gold become depleted. Start to let go of self-criticism and perfectionism and be kinder to yourself because once your pot of gold is brimming over, you will have some to share with others around you, including your children.

CHAPTER NINE

Our brains and nervous systems determine how we respond to our children

In the previous chapters, we explored the role of emotions and how the way in which we regulate our emotions shapes how we feel and react to people and events. We looked at the window of tolerance and how moving outside it, either through hyperarousal or hypoarousal, affects the way we deal with our emotions and in turn how others around us respond to us. As I outlined in the previous two chapters, the way a parent regulates emotion significantly shapes the attachment between a parent and child, the child's growing emotional regulation systems and sense of self, and also his or her behaviour in the moment.

In this chapter I'm going to outline how we come to feel defensive and tense or relaxed and calm around our children and how certain brain regions play a key part in this. I'm going to reveal how these brain–body systems change the way your body and mind respond to things and how this powerfully shapes so many aspects of the parenting process, including the connection between parent and child. Although there is a bit of brain science to take in do bear with me because this is so vital to building up a full picture of how connected parenting really impacts our children. To begin with, we need an understanding of the key brain players in the arena of emotional regulation, starting with the control centre for emotional responses, the limbic system.

THE LIMBIC SYSTEM: THE CORE OF THE PARENTING BRAIN

As a growing body of neuroscience shows us, parenting is shaped by emotional regulation pathways that reside mainly in the limbic system, the middle layer of the brain. It is the brain regions of the limbic system that make it possible for us to love and care deeply for our children, not least through the release of the powerful neuropeptide, oxytocin, which acts to increase our sensitivity to social signals, promotes bonding, trust when it is warranted, and that warm, intimate feeling you get when you're around someone you're really connecting with. Why does the limbic system have such a strong impact on parenting? To answer this, let's quickly explore what the limbic system does, what it is made up of and how those brain parts shape how we feel around our children.

What is the limbic system?

The limbic system, which you may think of as a hub for bodily based emotional regulation, is a collection of brain regions that work together and are densely connected upwards and downwards to various parts of the brain. It processes and amalgamates information coming up from your senses such as touch, sound, sight and smell, and from your internal bodily sensations, to which it then orchestrates a response. The limbic system processes the information we receive from our social interactions with others, such as their changing facial expressions. It makes an assessment of threat versus safety based on all these signals, and tells us whether we must open up to or close off to that experience. The limbic system structures involved in emotional regulation are more active in the right hemisphere than in the left and as I mentioned in Chapter Two and Chapter Three, this is because the right hemisphere is dominant in the processing and management of almost all emotions.

What does the limbic system do?

One of the primary concerns of the limbic system is to help us determine whether to approach or avoid things we encounter. When we are in 'approach' mode, we are inclined to be open to that experience be it physically, mentally or emotionally. When we are in 'avoid' mode, we close off in defensiveness to that experience, something that might manifest as simply as a tightening feeling in your chest and throat. This is linked to our nervous systems, and in particular our fight, flight or freeze response that originates in the most primitive layer of the brain, the brainstem, something I'll elaborate on later in this chapter. This 'opening up' or 'closing off' response happens at a bodily level in sometimes very small but significant changes such as your heart rate speeding up or slowing down, your breathing becoming shallow or deep, your tone of voice being soft or constricted and the muscles around your eyes tightening or relaxing.

Emotions play a key part in how the limbic system learns and is able to evaluate whether we must approach and avoid things because emotions provide a negative- or positive-feeling tone to our experiences. This information on the feeling tone of an experience is stored within the limbic system, much like a catalogue of emotional experiences, something that is referenced each time we encounter a similar experience in the future. This store not only shapes how we experience things but also how we feel in anticipation of them.

For example, if your child has had a monumental tantrum that triggered dysregulated emotions of anger, anxiety, embarrassment or confusion in you, the next time your child is in a situation that you think may lead to a tantrum, or is showing what you perceive to be 'tantrum warning' signals, you might automatically and unconsciously become tense, agitated and either hyper-aware of your child on the one hand, or closed off to noticing how your child is feeling and reacting in the moment on the other. Whether or not you realise it, this state of anticipatory agitation will cause changes in your face, body and behaviour that your child will sense, perhaps precipitating the very thing you fear – the tantrum! This is because, as you know, emotional states, especially at the level of the nervous system, are in constant communication with each other and therefore contagious. This is how the ongoing and subtle dance of momentarily opening or closing to experiences becomes a powerful mechanism that either facilitates or inhibits mutual connection between a parent and child.

The limbic system and the brainstem: Gut feeling, intuition and sensation

The limbic system has connections downwards into the brainstem, the most primitive layer of the brain, from which we receive sensory input from the body; from muscles, bones, organs such as the heart, intestines (the gut) and the lungs. These 'interoceptive' signals travel upwards via the spinal cord and the vagus nerve contained within it, into the brainstem, their first port of call in the brain. From here, the information is projected into the limbic system and the higher cortical areas, where, depending on whether or not you have developed a strong network of connections, it is processed and refined further, giving you the ability to contain and control emotions.

This information that originates in the body and is shared with the brain is often overlooked and ignored in our increasingly detached and cerebral way of relating to the world. It is these bodily sensations that constitute gut instinct, intuition and emotion and it is this information that is further processed by other brain regions in the right hemisphere to generate responses involving self-awareness, empathy and wisdom. The left hemisphere is active in controlling our behaviour according to social rules but this is different from doing the right thing out of feeling empathy or from an intuitive moral sense of right and wrong.

If we reflect on how we prioritise cerebral over sensory activities and visceral experience in both our own and our children's lives, we can see just how dangerous this trend can be over time. Our preoccupation with cleanliness is just one way in which we discourage children from connecting with their physical bodies, from touch, taste, nature and all those other experiences that bring the body and its sensations into our field of awareness. Our preoccupation with technology and its increasing command on our time is another example of this. We have moved so far from our natural embodied state that we now have to teach children, through mindfulness meditation, forest school and other socially constructed activities, how to tune in to the language of the body and sensation.

NEUROCEPTION, THE AMYGDALA AND THE DETECTION OF THREAT OR SAFETY

We know that our state of emotional activation fluctuates between high and low continually throughout the day to enable us to adapt and respond effectively to what we are experiencing. But how do we know what we are adapting to and therefore what the optimal response might be? The process of assessing whether we are emotionally and physically safe or under threat occurs through the process of neuroception[38] that takes place rapidly and outside of awareness in your brainstem, which is the most primitive section of your brain. This is not based on what you consciously know or see, but rather on the signals that your nervous system picks up through its monitoring of your inner and outer environment. To help you to get a balance between avoiding potential threats and approaching potential opportunities (for connection, food, flourishing), your nervous system must a) detect risk, and b) if you are safe, inhibit your fight, flight or freeze response. Without the ability to inhibit the primitive defence strategies of fight, flight or freeze, we simply wouldn't feel calm and relaxed enough to get close to people or to have successful relationships.

How does your amygdala help assess threat or safety?

This input from your nervous system is fed into other areas of your limbic system such as the amygdala, the insula and the temporal cortex (which responds to movements of the face and hands, voices). The amygdalae, commonly referred to in the singular as the amygdala, are two tiny almond-shaped structures buried deep in your limbic system that play a significant part in the parenting process, both in you and your children. Think of the amygdala as your brain's inbuilt alarm system, one that is hypersensitive to signs of threat, and that is always switched on. Information you take in and sense, whether external or internal, is processed by your amygdala for its 'quick and dirty' assessment of whether it is safe or threatening to you. If the amygdala could speak, you would hear it repeatedly asking and answering two key questions, 'Am I safe?' and, 'Are you with me?'

Once the amygdala has made its rapid assessment, it sends this information to other areas it is connected with, including higher brain regions in the cortex that allow us to reflect on and refine the original diagnosis, and where appropriate, to reassess or calm down our initial threat response.

Evidence suggests the brain, including the amygdala, leans more towards the detection and prioritisation of negative rather than positive information, lending the human brain a natural negative bias[39]. If you think about the threatening events these brain parts evolved for, such as detecting an angry or hungry predator, it makes sense that we are naturally more tuned in to sources of threat.

Memories, the amygdala and the hippocampus: our internal scripts

The answers to these questions, 'Am I safe?', 'Are you with me?' are not based on a rational analysis but rather on a non-verbal sense of things that will be shaped by your memories (emotional and factual), your general emotional regulation patterns, your attachment style and various other factors. The amygdala works with our stored memories to help refine its emotional assessment and add some context to our emotional experiences. Memories can be stored in implicit or explicit form. Implicit memories exist as a 'sense' of things, encoded as images or emotional tones, often things that we cannot put into words but that influence us in ways we may not recognise or make sense of. They are stored in the limbic system. Emotional memories, particularly those that evoked a strong emotion and those relating to fear, are stored in the amygdala and play a part in how we learn to be afraid of things. Explicit memories, on the other hand, are memories that exist as moments in time, usually events that we can recall and narrate, along with factual information such as what happened, when and where.

This is where the hippocampus comes in. The hippocampus, the brain's virtual filing system for explicit memories, is a close partner of the amygdala because they work together to determine whether things are safe or unsafe, based on our historical experiences. Typically, the hippocampus develops in the second year of life, usually from around eighteen months, prior to which your children will be encoding memories in implicit preverbal form in the amygdala and the limbic system. This is one of the mechanisms through which early experiences can shape us long after they occur, regardless of our ability to verbalise or recall them. The important point here for parents is to remember that just because children are too young to remember facts and experiences doesn't mean these experiences won't have an impact on them. The earliest experiences, particularly in the first few years of life, have a significant impact on later patterns of emotional regulation and attachment.

For example, if in your early years of childhood, you frequently experienced what you would have perceived as a rejection of your emotional needs and signals from a parent, either because he or she lacked the warmth to make a connection, or was preoccupied, unavailable, or simply wasn't able to read your emotional needs, you may find that you struggle to feel fully relaxed and comfortable when you engage with people. You may require a much higher baseline of safety before you can open up and feel emotionally comfortable. If you were shouted at as a child by an adult you perceived as threatening, you might find yourself tensing up or feeling nervous when you perceive any sign of aggression towards you or your child, even if that aggression isn't directed at you or you know, at a rational level, that you have nothing to be afraid of. You might even interpret actions as aggressive when they may not be intended that way. You probably won't be able to explain why this happens because you can't describe a particular episode that led to it.

THE DEFENSIVE STRATEGIES OF FIGHT, FLIGHT OR FREEZE

Your body and brain sense threat or safety before you consciously know it. You sometimes only come to know this when your emotional response becomes so heightened that you move out of your window of tolerance. If you have a neuroception of safety, you are within your window of tolerance and are able to relax and feel open to what is going on around you. When you have a neuroception of threat, you will experience the states of either hyperarousal, which takes you into the fight or flight mode, or hypoarousal, which results in the freeze mode, both of which are defensive states because they evolved to promote survival against real threats to our lives. Your body and mind will organise themselves for self-protection, which in relationship terms will result in a closing-off to people, and will set in motion a whole cascade of other brain–body events such as the switching on of your body's stress response.

We have only to think of the distress a very young child might experience in the presence of a stranger to understand how neuroception, and the threat and defence system, operate at a visceral and intrinsic level. The child has no conscious, cognitive awareness of why he is scared, and his parents may not understand why he starts to fuss and cry around someone whom they see as a perfectly normal, unthreatening human being, but his nervous system has registered a threat and he has no choice but to respond accordingly. Even if the child was older and was able to use a top-down higher brain process to modify his initial reaction, this would be secondary to his bodily based emotional response.

What determines which response will be utilised is unique to each individual. What might be a mild threat to one person and will switch on the fight or flight response, in another may be severe enough to necessitate a freeze response. A freeze response may occur when a person feels especially vulnerable, for example, being attacked or frightened by a larger person who has the ability to overpower them. It is the perceived level of vulnerability in the person facing the threat that determines how he or she will respond. The greater the level of vulnerability and the lower the level of control, the more likely we are to recruit the freeze strategy. Children, because they are vulnerable and dependent on us is so many ways, can feel deeply threatened and freeze in the presence of strangers, grown-ups who are not emotionally attuned, who struggle to regulate their emotions, or who hurt them physically or emotionally.

In healthy humans, we start from a baseline of safety and openness to social interaction. If that is not a viable strategy to deal with the situation in hand, we switch first to the fight or flight response to mobilise us to act against a source of threat. When the fight or flight response is inadequate to cope with the threat, we utilise the freeze response. As you read this, you may be tempted to imagine that these states are like boxes – you can either inhabit one or the other, but what I'm describing here is a shifting, fluid process that ebbs, flows and morphs in seconds. It is not a binary process because these states have open boundaries and we do not always know when one state shifts into another quite as clearly as this description might imply. What defines good emotional regulation is how responsively and quickly you are able to glide in and out of the various states associated with threat or safety depending on the demands of the situation in hand. This is what underpins connection and social relationships at all levels, including parenting, romantic relationships and friendships.

IS THE PROCESS OF THREAT DETECTION BALANCED?

Given this threat assessment system evolved and was honed in times when we encountered sudden threats from unexpected sources and we were also faced with a scarcity of resources, necessitating much violence and force against rivals to ensure our survival, neuroception is designed to occur rapidly with little time for calm, balanced analysis. This system evolved to help us detect tangible threats to our physical existence and yet, these days, it is activated mostly by threats that are psychosocial in nature; threats to our sense of selves; our status; our need for comfort, control and self-esteem. Being criticised can evoke a threat response, as can failing to meet a deadline or get somewhere on time, though these are not threats to our survival at all. We react as though they are because the brain mechanisms we use to deal with threats and stressors have not evolved and changed as quickly as our circumstances have. We are responding with old brains in a new world and because of this mismatch we sometimes find ourselves having monumental emotional responses to things that are really not threats in the true sense of the word.

Some of our amygdaloid reactions are unproductive and at times, downright damaging to our relationships. Every time a parent overreacts and shouts at their children, or become stressed and anxious around their children, they have entered a 'threat and defence' state. But we know, rationally, that children are not a threat to our physical existence. When we come to really think about it, what is so threatening about a child crying or complaining? What is so threatening about a child arguing and bickering with a sibling or even having a tantrum? In the long-term context, probably not much at all. But in the short term, we can end up at the mercy of an amygdala-driven threat reaction that temporarily short-circuits our ability to be rational, balanced and empathetic. It is not what the child might do, but rather our own and our children's emotions that we react to in a threatening or defensive way.

The amygdala develops early in humans, usually in the third trimester of pregnancy and can be functioning from birth[40] (Schore, 1994). Young children can very easily have a neuroception of threat because they are vulnerable and dependent on us for their survival. A parent leaving the room can elicit an assessment of threat as can a parent signalling disapproval by raising their voice or ignoring a child. Young children are mostly at the mercy of neuroceptive processes and the amygdala because they simply haven't developed the upper connections from the limbic system to higher brain regions that might allow them to reflect on, and rationally appraise, their own and others' reactions. This means they naturally demonstrate an intensity of emotion, often for reasons that we, with our higher brain regions, tend to dismiss, judge and feel impatient with. But, knowing what we do about the way the brain develops, we must hold in mind that they have no off, stop or reset button in their brains until much later, which means that we, as parents, must lend them ours until they develop their own. What this means in practice is that our primary role as parents is to first soothe our children when they are in the grip of an emotional response because they are not feeling safe. It is only when they feel safe once again that it makes sense to begin to correct them.

Neuroception can be faulty and the amygdala can be hypervigilant

Assuming your child is neurotypical, when we parent our children from a secure base, with a reasonable degree of attunement and emotional warmth, their system of neuroception works well, allowing them to detect risks where they may exist and feel safe when there is no threat present. In some people, neuroception may be perpetually faulty, resulting in someone either being blind to possible risks, for example when a child proactively engages with a stranger with no sense of wariness, or it might result in someone detecting threats where none exist, for example, interpreting someone's voice as 'angry' when it is neutral. When the threat-detection system is overactive, as it may be for people who have suffered trauma or have ASD, we spend too much time in defensive states and cannot feel 'safe' with other people at an intrinsic level. We might not like to get too close or be touched by people unless we are very comfortable with them, or we might find it difficult to really open up and trust people. At times, we may all suffer from faulty neuroception, such as when we are already under stress, are unwell or sleep-deprived.

Though the amygdala develops quickly in infancy, it is not something that develops in the same way in every human, irrespective of their experiences or genetics. If you imagine the amygdala as a volume button that can be dialled up or down, some people will have a naturally higher default setting, lending them a greater level of vigilance for threat and more of a tendency to react with anxiety or defensiveness to things that others may view as benign. For these people, though they cannot know it, their experiences are tinged with a suspicious, negative or fearful hue.

They are more likely to interpret and explain the behaviour of others around them in negative terms because the lens through which they first see the world, their amygdala, has developed in a way that fits the emotional environment of their early childhood or their gene codes.

The sensitivity of your own system of neuroception and your amygdala will have an influence on how relaxed or agitated you feel around your children. Fortunately, there are ways to calm and tame the amygdala, the most helpful of which can be the breathing exercises I have presented in this book. Compassion has also been found to have a soothing effect on the amygdala and this is especially relevant if you have children who are high in anxiety, are prone to emotional dysregulation or are under stress and pressure because they will be sensitive to shifts in your emotional tone and will need a higher level of soothing and regulation. I will show you an exercise designed to help you to cultivate compassion in this and the next chapter.

WHY WE LOSE OUR TEMPERS WITH OUR CHILDREN: THE LIMBIC SYSTEM HIJACK

Most of the description I have given you about emotions and emotional regulation in this book so far has focused on what we call the bottom-up system, which is based on the information you get from your body and environment – the physiological, non-verbal, intuitive sort of information your nervous system senses very quickly. The amygdala is part of this bottom-up world, where things happen out of our conscious awareness and speed trumps accuracy when it comes to processing information. In other words, when your bottom-up system is active, things feel as though they are happening to you and you have very little control over these gut reactions. It is so easy to become enmeshed in these sensations and run away with them into faulty thinking and overreactions.

If we think in sensory terms, this is the 'hot' system causing fiery and immediate responses. When this system is on, you are far more likely to take things personally, rather than being able to take a step back and think about the likely causes of the trigger situation. As an example of this system in action, think about the racing heartbeat you might experience if you hear a loud noise in the middle of the night; your first thought is likely to be that something is going to happen to you. For a few seconds, the worst possible scenario will race through your mind, before good sense prevails and you begin to think of more plausible explanations.

The top-down system is cooler and slower than the bottom-up system

The top-down system, on the other hand, is not directly connected with the body and receives input that has already been processed by the amygdala and other brain parts. It is a cooler system, based mainly in the prefrontal cortex (higher brain parts located behind the forehead) and helps you to be more detached from your immediate, instinct-driven emotional reactions. When this system is active, rather than feeling as though it is all happening to you, you are able to take a step back and ask yourself the equivalent of questions such as 'What is really going on here?' You can challenge and reappraise your immediate reactions and generate a sense of perspective, including noticing and perceiving the causes of things from another person's point of view. All input from the bottom-up and top-down streams ultimately ends up in the prefrontal cortex where a full assessment of information can take place in a balanced and integrated way, something that is vital to good emotional regulation, decision-making and self-control. As you will no doubt recognise, children really struggle with this in the early years. Everything is about them and rightly so, given they simply don't have the mechanisms to generate multiple alternative explanations for things.

When these systems are well connected and robust, certain brain regions that are closely interconnected with the amygdala – the anterior cingulate cortex (ACC), the orbitofrontal cortex (OFC) and higher parts of the prefrontal cortex – can physically inhibit the reactive, sometimes wildly unhelpful impulses stemming from it. This occurs when they send messages down to amend and temper the initial assessment of the amygdala, and in doing so they can alter our perception of events. This slower appraisal from the top-down system helps us to calm down, return to our window of tolerance, and adapt our responses. It allows us to inhibit our impulses to lash out at our children or run away from them, and to stay in control of our emotions and behaviour. It allows us to question, challenge and reappraise our initial assessment of a situation. This is the system that helps us control our initial urges, whether they be to eat a cake when we're on a diet, to shout at our children when we're busy and preoccupied, or to give in to desires that we know on some more cerebral level are simply not helpful or productive. Once again, bear in mind that all this takes time for children to master and in the interim, even though you may have told them fifteen times to put away their shoes, if there is something more compelling to do in that moment, they will find it hard to resist.

Though many of us will have developed a fairly robust control system, we all experience the odd 'hijack' when we lose our heads and act emotionally or impulsively in ways we later regret. This happens because the amygdala-driven bottom-up system is much faster than the top-down system. There are also more connections going up from the amygdala than there are coming down from the prefrontal areas, leaving us vulnerable to the impact of our initial reactions. The amygdala immediately has an impact on the prefrontal cortex but only receives a response back from the prefrontal cortex at a slightly later stage. If your amygdala has sounded a loud enough alarm bell, it can hijack your upper cortical areas and switch off sources of rational thought. This is much more likely to occur when we are stressed or sleep-deprived.

OUR THREAT SYSTEM EVOLVED TO PROTECT US FROM BEARS

If we think about it in evolutionary terms, and I mean all the millions of years during which the human brain has been evolving, the bottom-up system evolved in days where we faced frequent threats to our lives, for example, from predators such as bears. When you're trying to run from a bear you need to divert your energy reserves to action rather than thought. You really don't need to confuse yourself with a complicated analysis of which route to take through the trees as you run because a pause of even a second could mean instant death. You don't want to get distracted by pretty flowers, or the sun shining through the leaves as you run away, so your capacity for positivity diminishes. You also don't want to be thinking about other people who face a similar threat to you because empathy in such a situation might slow you down and leave you at risk.

So, your prefrontal cortex, with its capacity to reason, generate alternatives, solve problems and evaluate options is quite a liability in this context. This ability to turn off higher cognitive functions has adaptive value and persists to this day. When you are triggered by a loud enough amygdaloid alarm, regardless of whether it is a real threat to your life or simply a threat to your ego, sense of certainty or need for control, your ability to take other perspectives – in other words to have empathy and mental flexibility – is compromised, as is your capacity for problem-solving and feeling positive emotions. Your amygdala becomes even more sensitised to signs of threat, keeping you in a loop of negative and fearful reactivity. Typically, such a threat-based response would be short-lived and you'd recover your full faculties fairly quickly. But if you are more prone to emotional dysregulation it may take longer for your body and brain to turn off the threat signal. Also take note that stress and sleep disruption, especially persistent stress that you feel you have no control over, tends to lead to the disruption of top-down regulation and a heightening of amygdala reactivity. This is when we get stuck in a cycle of stress.

Sometimes our children are threats to us

You know by now that your amygdala sounds its alarm in a subjective way, so what is to one person innocuous noise from a child may, to another, be a source of distress. If, for example, you have been raised by parents who couldn't regulate their emotions, your children's emotions might be a source of threat to you and you may experience a limbic hijack simply by being around them when they are in the grip of strong emotions. If you have lots of internal rules about how children 'should' behave and your children are failing to live up to those standards, you will perceive what might simply be typical behaviour for a three-year-old child with an immature brain as 'bad' behaviour and will register this as a threat. If you're an introvert and need time to unwind and you are surrounded by noisy, talkative children, even something as innocent as your children talking to you when you're cooking or doing something else, might register as a threat. For some of us, it doesn't take much for a 'fight or flight' response to be triggered and unfortunately having children around seems to achieve this all too easily! To become less reactive to your children, you need to tame your amygdala and challenge some of your mental models about emotion and behaviour. All of the breathing exercises in this book will help you to achieve this with regular practice.

An exercise to generate a sense of safety, not threat, around your children

For this exercise, I'd like you to choose to do whatever form of slow, relaxed breathing that feels comfortable to you.

- As you exhale, slow down and relax, bring an image of your child's face into your mind.
- Notice any shifts in the state of your nervous system and regulate them so you are calm and relaxed.
- Pay attention to the left side of your face and notice if your mouth curls up (contempt) or the muscles of your face tighten in any way when you think of your child. The right hemisphere controls the left side of the body therefore certain threat-based emotions may be more readily expressed on the left side of the face. Soften and release tension in your face.
- Keep your exhale very slow and longer than your inhale and try to generate a sense of warmth (like golden light) around your heart.
- See if you can gently turn up the corners of your mouth into a tiny smile. This is associated with compassion.
- If you feel some stress or negativity when you think of your child,

that's perfectly okay. Without any judgement, tell yourself quietly in your mind, 'It's okay. It's just a child doing the best he/she can with the brain he/she has at this point in time.'

- Focus on pairing a lovely relaxation response with the image of your child.
- Try this with a few different images of your child until you feel calm and open.
- Repeat it with each of your children in mind.

Chapter Nine: Key points

- Connected parenting rests on the limbic system of the brain, which is like a hub for processing emotion-related signals from your body in response to what you experience, and which enables you to love and care deeply for your children.
- The process of detecting whether you are safe or under threat is undertaken by your nervous system and is called neuroception, something that occurs rapidly and outside of conscious awareness.
- The amygdala uses its own store of non-verbal, implicit emotional memories to assess threat. It also works with the hippocampus, which is your memory system for factual memories, for example, situations you can consciously recollect, to determine how to evaluate the experiences you face.
- You can only connect with your children when you are in a state of neurobiological safety, which is when you are in your window of tolerance.
- When you have a neuroception of threat, your amygdala, together with other brain parts, sounds its alarm and sets off a cascade of reactions that will include the defensive strategies of 'fight or flight' or 'freeze'.
- Children, because they don't have the higher brain regions to modify the initial threat assessment, have a lower threshold for entering defensive states. Because they are vulnerable, they can easily tip into hyperarousal or hypoarousal.

- The amygdala is part of the bottom-up system that is connected with bodily sensation; it is fast, reactive and fiery. Your top-down system is cooler, slower, detached from sensation, and able to evaluate things from a distance. When your amygdala triggers a loud enough alarm, it can hijack your top-down system, which will begin to shut down, trapping you in an emotionally reactive state that you struggle to calm down from.
- Sometimes our children can trigger a limbic system hijack that results in hyperarousal and dramatic reactions to things we later regret.
- Compassion quietens the amygdala and is a wonderful way of calming the threat-detection system. It can be generated through compassion-based breathing exercises and taking a wider, evolutionary perspective about how and why we all just find ourselves here, trying to do the best we can with the brains we have at this point in time, which are shaped by our genetics and early experiences, none of which are of our making.

CHAPTER TEN

Your nervous system underpins emotional regulation and connected parenting

Going back to the three essential conditions for a meaningful connection to emerge with our children, we covered open presence in the first few chapters, and then explored emotional regulation in subsequent chapters. In this chapter, we're going to delve into the third condition for connection, perhaps the most important in terms of how our children feel when they are with us, **emotional safety**. We've been through the process of neuroception and the role of the amygdala in assessing states of threat vs. safety. We know that connected parenting can only emerge from a neurobiological state of safety and that our ability to detect threat accurately and signal safety can be habitually or temporarily compromised, leading to defensive reactions in the absence of a real threat, or an inability to adequately assess risk and act appropriately in the face of potential danger.

In this chapter I'm going to outline to you how your nervous system plays a key role in heartfelt parenting, and how the way in which it shifts gear can either enhance or hinder connection. This is a really pivotal part of my approach to parenting and it rests on the wonderful calming properties of the vagus nerve, the upper part of which activates when we are in a state of safety and plays a vital part in facilitating emotional regulation, social engagement and intimacy. This is based on the remarkable work of Stephen Porges[41], whose work on the polyvagal theory has given us a profound understanding of how important psychological or emotional safety is to us on so many levels, including heart health, gut health and emotional and mental wellbeing. His work highlights how connection is a biological process that arises between people when we reciprocally regulate our inner states to come into synch with each other.

EMOTIONAL RECIPROCITY EQUALS SURVIVAL

As a race, human survival rests on the ability of a parent, usually the mother, at least in the very early days, to keep her child alive by feeding him and to nurture the child emotionally and physically until he is capable of meeting his own needs. She must want to care for her child, to notice what he needs, and to feel motivated to provide for those needs, however much discomfort she has to tolerate in the process. During vast swathes of our history where we faced great scarcity of food and other resources, this may even have meant forgoing food in order to safeguard the survival of our children.

For this to happen consistently and over the comparatively long period of 'nurture' required by humans before they are capable of full independence, we need to feel a great depth of love, pleasure and connection with our children, enough to compensate for the myriad challenges, sacrifices and difficulties along the way. Given how vital this is to our survival, it is no wonder that the limbic system is geared up to monitor and measure safety not just in physical terms but in terms of emotional connection.

In other words, safety is not just the absence of threat. To feel 'safe' we need to feel understood, cared for and connected with another human being on an emotional level because without this we would not survive. This is not about verbal connection but emotional reciprocity and synchrony in the moment. And because it is so important to our survival, our bodies and brains are designed to continually scan and monitor the faces of people around us, especially our primary caregivers, for signs of connection or disconnection.

Reciprocity is the basis of safety: The 'Still Face' experiment

When we initiate a social interaction, we have an inbuilt biological expectation of reciprocity, connection and coregulation, which, when violated, can generate a reaction of stress or anxiety. This may only last a second and the child may recover quickly, particularly if there are many moments of emotional synchrony and positivity in the relationship and the child has internalised an expectation of safety. But for some children the lack of reciprocity can be harmful and cause distress. When a child gazes at someone's face and doesn't get a sense of reciprocity (resonating facial expressions, matching, engagement), or encounters a face that is flat and devoid of emotional information, she can experience stress and confusion. Because very young children don't have a stable inner sense of themselves, not receiving feedback on their interactions may create additional anxiety. When you are too busy or preoccupied to really look at your children, it is distressing for them because social interaction and calm states are controlled by the same underlying system.

Edward Tronick[42, 43] designed a series of experiments to demonstrate how babies and toddlers respond to a lack of reciprocity from their caregivers. During the experiment, a baby and mother begin by interacting as they do normally for two to three minutes, with a reciprocal and mutually engaged level of interest in each other. The baby coos and the mother responds with a sound; the baby points and the mother automatically gazes in the direction he is pointing in. But in the next stage of the experiment, her face becomes flat, devoid of emotion, and she shows no interest in the baby's attempts to interact with her. The baby smiles, gurgles and coos but the mother's face remains expressionless. The baby tries hard to engage her, recruiting his best moves and sounds to entice her, but is met with unresponsiveness.

Although this only lasts for a couple of minutes, the baby grows distressed, looks away repeatedly, and finally starts to whine, then cry. The mother then re-engages with the baby and opens the channel to coregulation and safety, though this repair process is not always a smooth or an easy one either for mother or baby.

These experiments have been repeated several times and the results are consistent. During the still-face phase, the babies tend to avert their gaze, show more negative emotion and vocalise their distress through crying. They struggle to regulate their physiological responses and show agitation in their movements such as twisting and turning, and gesturing to be picked up. What this shows us is that the lack of emotional reciprocity and the unavailability of emotional information is stressful for children. Interestingly, in these experiments, the babies make gestures with their left hands to deal with the distress they feel. Bearing in mind the left side of the body is controlled by the right hemisphere and vice versa, researchers conclude that it is the baby's right hemisphere that processes the emotional upset caused by the lack of biological reciprocity.

DO YOU KNOW WHEN YOU ARE REVVING UP OR CALMING DOWN?

Your emotional state can enhance or hinder emotional reciprocity with your children and how well you are able to connect hinges on your nervous system. When your amygdala generates a threat response, it sends a signal to your hypothalamus to trigger your fight or flight response, which is enabled by your autonomic nervous system. Your autonomic nervous system regulates several bodily processes with a view to maintaining the equilibrium of these basic functions such as breathing, heart rate, perspiration, secretion of hormones from glands and so on. It needs to be able to allow you to dramatically increase the speed at which you use oxygen and metabolic resources for any type of action including fight or flight, but it also needs to restore you to a state of equilibrium so you don't wear yourself out and die from exertion or inadequate fuel. It facilitates this wonderful balancing act through its connections with the bodily organs, via the spinal cord and the brainstem. Most of the activity of your autonomic nervous system happens outside your conscious awareness, but interestingly you can exert an influence over it through relaxation methods such as consciously controlling your breathing.

There are two branches to the autonomic system: the sympathetic nervous system (SNS) and the parasympathetic nervous system (PNS). The sympathetic nervous system increases your heart rate and metabolic output and your parasympathetic nervous system lowers it. The SNS ramps you up for action whereas the PNS brings you back to a relaxed and steady state in your body. The way in which these systems work together in harmony plays a part in how well you regulate emotions, deal with stress, your physical health and your relationships.

The sympathetic nervous system – SNS

Your SNS enables you to prepare for action in response to information from, or about, your internal or external environment. It can do this to prepare you to defend yourself against a threat by enabling your fight or flight response, or when you are engaged in purposeful physical or mental activity such as exercise, playful behaviour or dealing with demanding chores, deadlines, or challenges at work. It is an active state, during which the chemicals adrenaline and noradrenaline are secreted. The release of adrenaline prepares us for action via increases in heart rate, changes to breathing patterns, enlarging of the pupils of the eyes, energy release and mental alertness.

When your fight or flight mechanism is on, your stress system, the HPA axis, is also switched on, which leads to the stress hormone cortisol being released. In the short term, cortisol release can be helpful in reacting to challenging situations but when you are unable to calm down, either because the source of the stress is ever-present, or you are unable to regulate your emotions well enough, you get stuck in a cycle of stress. Long-term stress and high circulating cortisol levels can reduce your 'feel good' hormones dopamine and serotonin, leaving you unable to derive pleasure and contentment and at greater risk of anxiety and depression.

Persistent cortisol levels also lead to inflammation and affect your heart health, resulting in an increased risk of heart disease, hypertension, increased blood pressure and stroke. It can cause sexual dysfunction, weight gain, hormonal imbalances and muscular tension. It impairs the functioning of your hippocampus, your memory system, resulting in memory loss and further disruption to your emotional regulation capability. Bearing in mind that the SNS can be activated by anything you perceive to be a threat or a demand on your energy and resources, including chores, exercise, emails that bother you, negative or distressing thoughts, challenging experiences, and every time your child interrupts you to ask for things when you already have much on your mind, is it any wonder that modern life leaves so many of us feeling frazzled, ramped up and stuck in a mode where we struggle to switch off? And what impact does this have on the inner state of our children?

The parasympathetic nervous system – PNS

The parasympathetic system is responsible for calming your bodily reactions and bringing you back to your baseline state of equilibrium, where your heart rate is slower, you are conserving energy, able to digest your food, and you feel calm, content and relaxed. It is your PNS that is dominant whilst you are sleeping as opposed to your SNS that is activated when you are waking up in the morning. Because of this the PNS is often referred to as the 'rest and digest' system. Most exercises that aim to trigger a relaxation response, such as meditation and yoga, do so via long exhalations because they stimulate the PNS, which calms you down.

When the PNS is active, you are more aware and responsive to information from within your body and you're able to switch off from your outer environment. The neurotransmitter acetylcholine allows your heart rate to calm down and your internal organs to restore themselves. Your SNS and PNS are designed to work together in synchrony moment by moment, but our lives are so fast and full that the SNS can start to dominate our experiences, upsetting the natural balance, leaving us unable to calm down enough to really connect with each other.

THE PARASYMPATHETIC NERVOUS SYSTEM IS UNDERPINNED BY THE VAGUS NERVE

The vagus nerve is one of the primary nerves of the parasympathetic nervous system. It's a long, wandering nerve that begins in the brainstem, at the base of your head, and travels down through the spinal cord where its many bundles of nerves extend into organs such as your heart, lungs, stomach, intestines, kidneys and genitals. It is a very busy nerve that conveys signals from the body up into the brainstem, from where these signals are propelled upwards into the amygdala, insula (which facilitates empathy and bodily awareness) and orbitofrontal cortex (which helps us process social and emotional signals from other people), and further up into the prefrontal areas for more sophisticated processing. It also acts as a conduit for signals down from the brain to the face, head and organs to modify their state in response to the brain's assessment of safety versus threat. But it's not just a simple communication cable – it does much more than that.

The vagus nerve has two branches, the lower vagal system and the upper vagal system, each with a different function.

The lower vagal system is responsible for the freeze response and digestion

The lower branch of the vagus nerve, the lower vagal system, is a more primitive system that connects the organs below the diaphragm, such as the stomach and intestines, to the brain and carries messages from the internal environment of the gut upwards, helping regulate digestion and relaxation. It is prominent when we are relaxing, content in the moment, sleeping, and holding our breaths. When this system is in the lead, there is little motivation to do anything except be in the moment. When we are in a state of safety, we use the lower vagal system to repair and restore our cells and organs, which we cannot do when we are in a state of threat. When we are in the state of threat, we can recruit the SNS for a fight or flight response. But what about the freeze response? It is the lower vagal system that orchestrates the ancient and primitive freeze response of

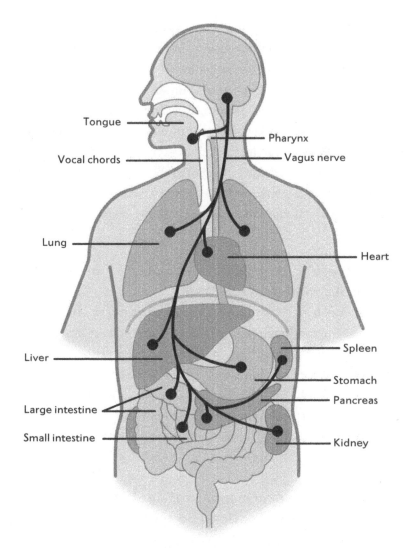

The Vagus Nerve

immobilisation in response to a threat, where your heart rate and levels of energy drop very low and you are in 'shut down' mode. This can occur momentarily, such as when you feel frozen like a rabbit caught in the headlights, or it can happen over a longer period of time. It can also vary in intensity.

When your SNS is active, this lower vagal circuit is suppressed so that we can sideline digestion and relaxation, and focus energy on dealing with challenges and threats. People who are emotionally dysregulated or under stress and are prone to either hyper- or hypoarousal may suffer from gut related issues such as Irritable Bowel Syndrome because they aren't able to utilise the lower vagal system effectively. I have seen this anecdotally in my work with corporate clients; those of them who habitually suppress emotions rather than expressing and dealing with them openly also report greater symptoms of gut discomfort. You might also notice from your own experience, and from observing your children, that gut discomfort frequently coincides with periods of emotional dysregulation.

The upper vagal system helps us to regulate emotion and remain in our window of tolerance

The upper branch of the vagus nerve, the upper vagal system, connects the organs above the diaphragm such as the heart and lungs to the brain, enabling us to keep our heart rate and breathing in a healthy range that supports effective connection rather than defensiveness. It is like the conductor of an orchestra, coordinating, controlling and limiting the SNS and the lower vagus to allow us to stay calm and present. The upper vagal system connects with the sinoatrial node on the heart that modulates your heart rate, and acts as a 'vagal brake', keeping your heart rate within a healthy ambit and allowing you to stay in your window of tolerance. Without the vagal brake, our hearts would beat much faster and we would struggle to remain calm enough to manage relationships well. We would probably be in either fight or flight or freeze mode all the time, none of which are able to support nurturing behaviour.

Because the upper vagal system evolved later, it has the ability to control and inhibit the older systems of the SNS and the lower vagal branch of the PNS, and prevent you from entering the fight, flight or freeze modes when these states might be an inappropriate response to the situation in hand. This is a clever system because simply by easing its foot on the vagal brake (your heart's pacemaker) a little, it can allow us to rapidly and smoothly increase heart rate to deal with situations that demand it but that don't require as strong a response as the sympathetic nervous system (SNS) activation, which is quite taxing on the body and a bit slower to respond than the upper vagal system. Similarly, it can press the brake down and enable us to lower our heart rate and calm down a little without having to go into the freeze mode, which involves shutting down and would be too extreme a response in most of the scenarios we humans encounter. To understand how fluidly this system works, imagine you're on the move. When you're walking at a reasonable pace, the upper vagal system can manage this quite well just by slightly releasing the vagal brake, but when you start to run, the vagal brake turns off to allow the SNS to take over because running requires a much higher level of energy and therefore a higher heart rate.

Communication is a perfect example of this system in action: when you are speaking you need a rapid but slight increase in heart rate and bodily activation, which is enabled when the vagal brake relaxes its control of your heart rate. In contrast, when you are listening, you must calm and slow down so that you can take in the other person's response to what you have just said without interrupting, talking over someone or getting so stimulated that you can't participate in the conversation in a reciprocal manner. If your heart rate is too high because you have tipped out of your window of tolerance into hyperarousal, you might become too excitable or threatened to listen well. The immaturity of a child's upper vagal system can explain why children are so prone to interrupting and talking over each other when they are excited, eager to get a point across or anxious. Children aside, I often see this in grown-ups too, particularly in work or social situations.

It is the vagal brake with its ability to ramp up just enough or slow down just enough that allows us to respond to people and challenging situations from within our window of tolerance. This is at the heart of resilience. The upper vagal system enables us to remain calm, mindful and empathetic, even in the midst of doing several things at once, when we have multiple demands made of us, and when our children are dysregulated or resistant. We all know how tough that can be but some of us have an easier time of it than others depending on how well this upper vagal system works. When your upper vagal circuit is in charge, your fight or flight response is turned off, your stress response is inhibited, you can authentically connect with people and your body can turn its attention to healing and recovering from stress and strain. It is nothing short of fundamental to wellbeing and emotional resilience.

The upper vagal system enables heartfelt connection without threat

Over the course of our evolution, the part of the brainstem that the upper vagal system originates in, the nucleus ambiguus, has also become the part of the brain that controls certain muscles of the face, leading to a face–heart connection. Not only does this upper vagal circuit connect with the orbitofrontal cortex and other brain parts that process and respond to emotional and social signals, but it also controls the striated muscles of the upper face, for example, the tiny muscles around the eyes that convey emotion, the middle ear muscles that allow us to tune in to the sound frequencies of human speech, certain muscles of the head and neck that allow us to move our faces when we are speaking and also our larynx and pharynx that modulate the musicality of our voices when we are speaking. It even plays a part in swallowing, which is why we find it hard to swallow when we are in the grip of powerful emotions. It is these features of the face and voice that provide the neuroceptive system and the amygdala with the information they use to assess threat and safety levels in our interactions with people. These muscles in our faces are the conduit through which our emotions bring us to life, through our head movements, our gaze, our eye movements and importantly, our tone of voice. All these elements in tandem make up what Stephen Porges calls our **social engagement system**.

When the social engagement system (upper vagal circuit) is switched on, which can only happen when we have a neuroception of safety, our heart rate shifts flexibly within a range that supports the ebb and flow of social interaction. Our facial muscles relax and become more mobile, allowing us to express a wider range of emotions, and we are able to make gentle yet lingering eye contact, rather than appearing flat, fixed in a stare, too intense, or unable to look someone in the eye beyond a fleeting glance. Our voices naturally become softer and more musical, demonstrating a greater range in pitch and tone. We are prone to spontaneous smiling rather than doing it because we know it is socially desirable. We are able to listen well, picking up not just on the words but on the shifting emotional states and intentions of the speaker.

When the social engagement system is temporarily turned off or not developed enough, the upper part of the face is less mobile and expressive, whereas the lower part of the face (which is more connected with the SNS) might become more active. When we have a neuroception of threat, our eye gaze changes as does the frequency and depth of eye contact, the tiny muscles around the eyes become frozen or tense, the tone of voice becomes duller or sharper, we might end up 'barking' at people or talking at them, and we struggle to block out external background noises and tune in to human speech. This is not a state that can support empathy and connection with our children because we cannot read their inner states especially well, and we cannot convey a sense of safety to them. People who have not developed a strong social engagement system, or who have tipped temporarily into threat and defence mode, may talk at length without recognising the other person's need for reciprocal communication; for give and take. Because the same system that allows us to pick up on the tone of a human voice is also the system that regulates our heart rate, our heart rate can rise and fall according to what we hear and see in our interactions with people with whom we expect or need close social interaction.

Safety, connection and physical health in your children is shaped by your social engagement system

Safety emerges when there is no neuroception of threat and we feel connected through warm and attuned social interaction. During any state of threat, mild or severe, the sympathetic nervous system or the freeze response of the lower vagal system will take over, and the upper vagal system will switch off its role in connection, digestion, relaxation, bodily repair and restoration, and other health-supporting activities. The system that allows us to regulate heart rate and breathing – and to feel calm, safe, emotionally regulated and physically healthy – is also the system that allows us to express and read emotions on people's faces.

Emotional safety, through interacting with people whose social engagement systems are switched on and working well, is inextricably linked with calming, soothing, physical health and wellbeing, making reciprocity, kindness and heartfelt connection no less than a fundamental human need. This completely transforms our understanding of what children need from us and from other adults with whom they interact. Children don't know it, but their brains are constantly scanning your face and voice to assess where they stand with you. Will you show them the reciprocity they need in order to feel safe and supported and can they trust you? Or must they close off to you because you might knowingly or unknowingly reject or frighten them? Without knowing it, you reveal your inner state through the tiniest of changes in your face and voice.

Children feel safe when they are interacting with people who make gentle eye contact, have expressive faces and soft musical voices because this is what promotes the neuroception of safety. Mothers automatically engage with their babies using a distinctive pattern of speech and facial expression. Facial expressions are exaggerated and we adopt a particular pitch and tone of voice. Research finds this is universal and emerges spontaneously – we don't need to be taught it because it is what our right hemispheres intuitively know how to do. And it is not only what babies respond best to (they prefer it over adult-like speech) but also what their brains need in order to grow the circuitry for healthy emotional understanding and expression. This wonderful choreography of facial expression and body language between mother and baby is made possible by our social engagement systems. It tells your baby that he is safe, seen, understood, and will survive.

It is no accident that grown-ups automatically speak to babies in soft, high-pitched voices (called motherese) but the result of an intuitive, biological understanding of how to soothe and calm the inner state of an infant. We humans are hardwired to tune into sounds: low-frequency sounds (machines) can trigger a neuroception of threat, as can very high-frequency sounds. It is the frequency of human speech that emerges from a healthy social engagement system that is most soothing to us. This comes from acoustic stimulation that is most similar to the sound frequency of a mother singing her baby a lullaby. Please note that some of us have neuroceptive systems that might not register human speech well or might be extraordinarily sensitive to the tone of the human voice and therefore not easily soothed by it.

When you are busy or stressed, and your SNS is in the lead, your social engagement system will switch off, creating a reciprocal state of stress or vulnerability in your children. If this occurs regularly, and your children don't have time to connect with you in a relaxed, slow and restful state, it will have an impact on their emotional and physical wellbeing. Whilst we don't need to be in a state of safety all the time, we, and our children, do need it at least some of the time. Take a moment to slow down your heart rate by pairing a long, slow exhale with the softening of all the tiny muscles around your eyes and jaw, and really focus on them when they speak to you, even for a few moments. Try to let yourself go completely in that moment and just be there, immersed in that feeling of being in synch with your child.

An exercise to downregulate the SNS and activate the PNS

The aim of this exercise is to tone down the SNS and re-engage your upper and lower vagal circuits. If you're judging and demanding something of yourself, you'll be activating your SNS, even if it is for the goal of activating your PNS! So, just let go of striving and adopt a mindset of curiosity rather than judgement. There is no right or wrong way to do this; anything is better than nothing. Rather than judge and evaluate, just notice. In all these exercises, try to tune in at a bodily level with what you *feel* in your body, not what you think. This will be very hard for many of you but do give it a go. I have had clients who could not notice their heart rates at all when they first started these exercises but were able to after several weeks of practice.

Sit down somewhere quiet and with your body relaxed yet in a dignified upright posture, close your eyes and follow the steps below.

1. Start by trying to regulate the inhale and exhale so they are of the same duration, for example, four counts in and four counts out. As always, inhale up into the space behind the middle of your forehead and exhale out into your heart and chest area.

2. As you exhale to the count of four, soften your exhale and try to relax the muscles of your face and upper body.

3. After a few minutes, start to lengthen your exhale slowly to the count of five or six.

4. Pause for a count of four between your inhale and exhale if you can comfortably do that. Pausing activates the lower vagal circuit in a state of safety.

5. Inhale fully, then try to constrict your throat slightly and make a sound as the air travels down your windpipe during the exhale (this is the 'ujjayi' breathing practiced during yoga). This type of breathing calms and balances your nervous system; if you persist with it for a few minutes, you will notice that your mind becomes clearer and you will palpably feel yourself slowing down. You don't have to do the 'ujjayi' breath if you don't want to – a regular exhale will work just fine.

If you notice you are stressed, agitated or simply too mindless to do this exercise, you might need to take a few deep inhales and sharp exhales. Release the exhale however it feels comfortable (this might be quite a forceful exhalation – that is fine) and let your body slump a little. Repeat this a few times until you feel you have reset yourself. Then try to do the full exercise as described here.

Chapter Ten: Key points

- Babies are biologically hardwired to monitor the emotional responsiveness, or reciprocity, of their mothers to their attempts to engage with her. The level of synchronised, reciprocal emotion between the parent and the child defines how safe the child feels.
- When your amygdala detects threat or safety it transmits a signal to your autonomic nervous system to respond accordingly.
- The autonomic nervous system has two main branches: the sympathetic nervous system (SNS) and the parasympathetic nervous system (PNS). The SNS readies you up for action and the PNS restores and calms you. These shifts are not binary, but rather more like the changing colours of a chameleon – there are different colours and gradations in shades of colours depending on the background situation.
- The PNS is underpinned by the vagus nerve, of which there are two branches, the upper and the lower vagus. The lower vagus enables the freeze response when we are under threat and when we are in a state of safety, and it enables rest, digestion, organ and cellular growth and restoration.
- The upper vagal circuit is active when we are in the state of safety and within our window of tolerance. The vagal brake enables us to shift our inner state in response to what we might be saying or hearing but to stay regulated and calm overall.
- A profound point to note about this system is that the brain part this upper vagal circuit originates in, the nucleus ambiguus in the brainstem, is the same part of the brain that controls certain muscles of the face, including muscles around the eyes that we use to convey emotion, the inner middle ear muscles that allow us to tune in to human speech over background noises, the musicality and tone of our voices, and also the muscles of the head and neck that enable us to move them when we are communicating.
- The heart connects directly with the human face, the portal through which we express love and emotion, giving heartfelt parenting a different layer of meaning.

- When you are not in a state of safety, your child will read it on your face and in your voice, and will know it through the lack of emotional reciprocity. This could catapult him into an SNS reaction, however mild or strong, because he is unable to have a neuroception of safety.
- The upper and lower vagal systems of the PNS together are responsible for sleep, digestion, organ and cellular repair and growth, emotional regulation, emotional expression and human connection, all of which can only take place in a state of safety. It is one system that connects all these elements.

CHAPTER ELEVEN

How does emotional safety influence your day-to-day parenting style?

In the previous chapter I described the branches of the autonomic nervous system, the sympathetic and the parasympathetic nervous systems (SNS and PNS). I also outlined how the upper vagal system of the PNS acts as a vagal brake, allowing us to regulate emotions flexibly by moving in and out of states of increasing activation and calming on a moment-by-moment basis.

In this chapter, I'm going to bring to life how these systems of the nervous system have an impact on our day-to-day interactions with our children. I'm going to help you to notice and manage your social engagement system so that you can parent in a way that feels enjoyable to both you and your children.

HOW THE SOCIAL ENGAGEMENT SYSTEM ENABLES DIFFERENT STATES RELEVANT TO PARENTING

The states of threat or safety have an impact on the way your nervous system shifts gear. Rather than thinking of these shifts between the sympathetic (upregulation) or parasympathetic (downregulation) nervous systems as an either/or binary process – as though you can either be ramped up or calm but nothing in between – imagine yourself as a chameleon that can rapidly and suddenly change colour to match the background. You're a sophisticated chameleon so there are various shades of colour you can adopt, and the shades of colour are graded from mild to moderate to very bright or dark. This is how your nervous system works, ramping up or slowing down in degrees depending on what you need to do to best respond to what is going on around you.

What defines good emotional regulation is how responsively and quickly you glide in and out of the various states depending on the demands of the situation in hand. There are several states that are enabled by the SNS and the two vagal circuits of the PNS, some of which are very relevant to what our children need from us. You can see from the diagram how the level of activation combines with the degree of threat or safety to produce various states that will have an impact on parenting.

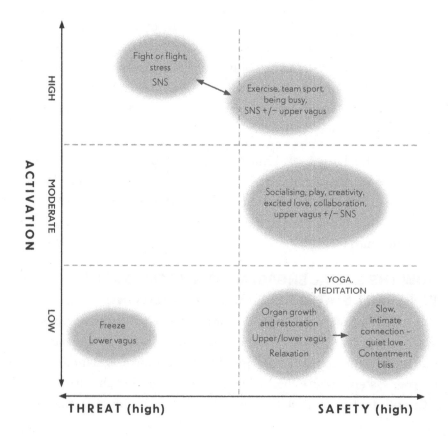

Being playful and joyful exercises the social engagement system

Being playful and joyful with our children is an important facet of parenting and forms the basis of 'excited love'. Experiencing mutual pleasure, happiness and excitement through engaging with our children builds their brains for creativity and lifelong positivity. Active play is like a workout for the social engagement system because it can involve increasing levels of mobilisation, involving either the vagal brake or the SNS (heart rate increases), in the absence of a threat response. Team sports that involve SNS arousal, and also require social engagement, are a good example of how the SNS can be coactivated with the upper vagal system to maintain a mode of engagement that is productive in the situation. Doing physical activities with our children that support social engagement and having fun whilst using the SNS is a good way to build their capacity for regulation. But do bear in mind that when we are already in a stressed state this can backfire because the social engagement system will be dampened, and greater SNS arousal when we are already stressed will simply lower the threshold for an emotional fight or flight flip-out!

Children, especially those who engage in playfighting and other forms of 'horseplay', sometimes tip quite easily from friendly to aggressive because their social engagement systems haven't matured enough to keep them regulated when they experience rapidly increasing levels of physiological activation. They are also more liable to take things personally and react defensively because their higher brain circuits are still in development. What helps children and adults stay regulated in a state of play or very animated social interaction is how well they read facial expressions. It is through making eye contact and reading our playfellows' faces that we infer their inner state and therefore come to sense whether we are still playing or whether it has altered into something we need to be wary of, or try to defend ourselves from. Children often return to a state of play or social engagement fairly quickly and parents must be cautious about interfering with what is mostly a natural and intuitive process between human beings. However, children whose social engagement systems are underdeveloped or compromised will struggle to notice, read and process these signals from others' faces.

The social engagement system allows us to experience still, quiet intimacy

When you have a neuroception of safety and your SNS is dialled down, your upper vagal circuit can simultaneously coactivate with the lower vagal circuit (freeze) to produce intensely still, calming and intimate states of connection. This is 'quiet love'. Think of it as momentarily freezing but out of pleasure rather than threat. It is a deeper, slower kind of connection that produces a loving feeling, for example when cuddling, that is gentle and slow, leaving us awash with oxytocin. This system can be activated during childbirth, breastfeeding, during sexual intimacy and also in moments of emotional resonance with someone whom we are close to. Sometimes, it may involve something as simple as a lingering moment of eye contact, or a shared smile, but the impact of the moment far outweighs its duration. This feeling is the bedrock of intimacy in relationships though it can happen at any time with anyone, even a stranger or a friend. It is unlikely to occur in those of us who are avoidant and tend to suppress emotions. Some parents, depending on their early experiences, may struggle to experience such states of intimacy with anyone, including their children. But many of us can train ourselves to move closer towards this state and it really is worth the effort, in more ways than one.

How do you know when this system is in the lead? The primary indicator is that it feels marvellously soothing, relaxing and restorative. Even if it only lasts a few moments, you'll feel a sense of stillness, ease and contentment and an absence of a sense of striving or judging. When you go into this mode, however briefly, your tone of voice will become more soft, calm and melodic. You'll find yourself speaking more slowly or not at all, and naturally pausing more, which helps you to notice the reaction of the listener. Your face will soften and all the tiny muscles around your eyes and mouth will relax and uncrease. You can look into someone's eyes and feel moved by what you sense. Your body will release some of its tension, and your mind will be open and present in the moment. This has a powerful calming effect on children, and even on grown-ups, because it triggers a deep feeling of safety and a surge of the hormone oxytocin. When you touch your children in this state, it calms their nervous systems, helps them rest and relax, and nourishes the cells in their bodies and brains.

Are we becoming imbalanced in the way we move between nervous-system states?

Although we can switch rapidly between SNS and PNS activation on a moment-by-moment basis, there does need to be a balance for optimal wellbeing and connection. The upper vagal system of the PNS is our baseline from which we catapult away into SNS activation to deal with challenges, movement and all the things we need to do. These days, it seems the balance is skewed with many of us spending too much time in active SNS states and not enough time in the slower, calming states of the upper and lower vagal systems together. We are becoming increasingly individualistic in the way we live our lives and are finding caring, nurturing and connecting with others stressful rather than restorative. When we are responsible for caring for others but we cannot activate a state of safety from being around them, we become stressed and resentful. Needing or wanting to get away from your children might be a sign that you are unable to experience being around them as intrinsically rewarding.

HOW WELL DOES YOUR CHILD USE HIS OR HER SOCIAL ENGAGEMENT SYSTEM?

Not all children have the same capacity for social and emotional regulation as there will naturally be variations in underlying brain–body systems that make up the social engagement system. Some babies will need a higher level of soothing and coregulation than others so parents need to adapt and respond in a flexible way depending on what their children can tolerate. Some children have a low threshold into SNS activation and can become easily overstimulated, agitated, excitable or anxious. When engaging with your children, pay attention to how their social engagement system works. Are they generally able to listen well or do you often need to repeat yourself before they can take in what you are saying? Do they seem to react quickly to changes in your tone of voice, or are they oblivious? Are they prone to freezing? Do their facial expressions and tones of voice vary whilst they speak? Or are their faces relatively flat (this can look as though they are calm). When they look at you and they are describing something that elicits an emotion in them, does the emotion show in their eyes? These are all signs that their social engagement systems are either working well or are compromised, either temporarily or habitually.

Some children, for example those with Autism Spectrum Disorder (ASD), may have an impaired ability to read emotional cues in faces and to infer the emotional states or thoughts of others. They may be compromised in their ability to exhibit spontaneous and reciprocal eye contact or communication. They often demonstrate auditory sensitivities and struggle to tune out low-frequency background sounds and selectively tune in to and comprehend the human voice. Their capacity to regulate emotions is affected along with their inner 'sense' of what is and isn't appropriate in a social context. They tend to have a high level of amygdala activity triggering higher than typical levels of anxiety and may need a higher level of soothing, or even a different type of interaction, to feel safe and relaxed.

If your child is prone to dysregulation, what you need to do depends on the trigger. If the child is generally overstimulated, take a look at yourself and notice if you are in a calm inner state. Coregulate your child by activating your social engagement system through a quick breathing exercise: in for four, pause for four, out for five, paired with an image of your child smiling and happy. This can be done very quickly but if you are already in fight, flight or freeze mode yourself it may take longer. Because we are so hardwired to scan what is going on around us for cues as to whether we are safe or under threat, too many sources of visual or auditory stimulation can be overwhelming for children and lead to hypervigilance. Every child is different but man-made sources of stimulation are not necessarily positive for our children, especially when they are young and haven't learnt to desensitise themselves to these things. You might want to look at your child's sleeping patterns and whether your lifestyle is promoting enough of a balance between things that activate the SNS, creating a lower threshold for stress or anxiety, and things that can calm and soothe the child. Prioritise sleep and rest for a while and try to have at least one day a week for the whole family to slow down without any feelings of pressure. On this day, try to relax and let the children have time when you are not giving them instructions on what to do and just allow them time to be.

EYE CONTACT, THE AMYGDALA AND PARENTING: DO WE REALLY CONNECT?

The amygdala and other limbic areas involved in neuroception are especially sensitive to shifts in gaze, alterations in the size and depth of light in the pupils, and micro-movements of the muscles around the eyebrows and the wider region around the eyes. These facial signals also play a key part in empathy, the ability to understand what other people may be thinking and feeling. If we take a moment to think about the implications of this on parenting, and especially parenting in this age of being busy and mentally absorbed in other things, it is concerning that one of the consequences of this is compromised eye contact. If children make an assessment of safety based on eye contact with their parents, and being busy and preoccupied reduces opportunities for genuine eye contact, this leaves them feeling disconnected and uncertain about their level of emotional safety around their parents and in the wider world in general. It also makes it harder for them to assess their own experiences because so much of their assessment initially depends on how they observe their parents responding to things.

As I laid out in the chapter on why attachment matters, your child develops his sense of himself through the non-verbal emotional feedback he receives from you. As children grow older, they internalise a view of themselves as generally safe with people; violations in eye contact may not impact them so much because they have a store of positive memories of social interaction to draw on. But most humans, whether children or adults, will experience an inner shift when there is a violation of biological reciprocity through body language with people whom they are close to and look to for comfort and connection.

A few years ago, when I began studying the science behind emotional regulation, I noticed that when I was too engrossed in things I had to do, or generally busy and stressed, I tended to talk 'at the children' without really slowing down enough to look properly into their eyes. I'd rather they just listened to me and got on with things so that I was free to do what I needed to do. I learnt that this disconnected way of communicating with them, if it persisted for a certain length of time, inevitably led to some form of dysregulated behaviour in my children, sometimes a slight detachment from me or on the other hand, excitability. I notice it even now. As soon as I go through a period of a few days where I have persistent things on my mind that take me away from being present in the moment with them, my younger child becomes increasingly dysregulated and easily overstimulated, which is in stark contrast to her usual regulated manner. Take a moment to think about what happens to your level of eye contact when you are busy or stressed. Do you really even 'see' your children and their inner state?

What does your face convey to children?

Children are very tuned in to whether facial expressions are dismissive, hostile or warm and accepting. When you look at your children with negativity or judgement, or perhaps avoid eye contact from a place of dissociation or avoidance, this will be a source of stress to them. When your child neurocepts a negative expression in your eyes or observes your facial muscles crinkle up in disgust or anger, even for a very short duration, he will make an assessment of threat, as a consequence of which he will become preoccupied with self-defence and trying to self-regulate his emotional response, a state incompatible with open-mindedness or listening. Of course, this can dissipate as quickly as it arises but if your children have a repeated sense of not feeling 'safe' around you, it will leave its mark. There is a compelling body of evidence that tells us that negative experiences leave a longer and more influential trace than positive experiences in relationships[44], which means we have to work hard to repair negative emotional experiences with our children.

If this happens on a regular basis, your child may become either hypersensitised to these signs of threat (hyperarousal) and might come to anticipate negative reactions from others habitually, or he may become desensitised to them (hypoarousal/dissociation) and close himself off to them. On the other hand, if your relationship with your child is generally secure, these temporary displays of negative emotion will not be overly threatening to your child because there is a general foundation of trust and safety, and your child knows things will soon be repaired.

Given eye contact plays a part in empathy, we must try to look kindly at our children when they are dysregulated, even if we are creating the distress through our own interactions with them. This requires self-awareness and self-regulation skills; firstly, to notice whether you are present or not, and secondly, if you are not, to regulate yourself to a state where you are. When we are able to harness our own powers of self-awareness to activate our social engagement systems with our children – and this can be done intentionally and with just a small degree of regular practice – it has a disproportionately positive impact on your child's emotional regulation capabilities because it signals that your child is safe. It is only when a child feels safe that she can properly attend to the important business of exploring, creating, playing, connecting and being joyful. Please know I am not talking about left-brain-driven conscious eye contact or arranging your face in what you think is representative of social engagement, which will look odd and confusing to your child, but something that will emerge naturally when you learn to cultivate a truly relaxed, open state in your body.

WORKING WITH YOUR SOCIAL ENGAGEMENT SYSTEM DAY TO DAY

Try to pay attention over the next few days to the quality of your interaction with your children and particularly whether you feel relaxed and soft around your face, heart and chest, or tight and constricted. The most effective way to turn on your calm and connect system is to slow down your breathing, pause, talk less and simultaneously soften and relax your body. The tiny muscles of your face might scrunch up when you are preoccupied, stressed or annoyed so you might need to think about relaxing them. Notice the tone of voice you adopt with your child, and how often it ranges from tight and sharp to angry, however small the increase. How often is it soft and gentle, as you might speak to someone when you are vulnerable or really content in the moment? Reflect on how well your child can be expected to respond to you when her brain moves from an open state to a defensive state based on what she has taken in from your face and voice.

You know from how the upper vagal system connects with the inner ear muscles that the less safe she feels, the less she can process what you are trying to say. Some children are especially sensitive to shifts in your tone of voice and can easily flip into agitation, anxiety or even shame when your social engagement has temporarily switched off. It is important to lower your volume and soften your tone of voice if your child is stressed, anxious or you are about to do something that might evoke a 'shame' response, like telling them off for something. Yes, I know this sounds almost impossible and the opposite of what we typically do, which is to raise our voices when we chastise them. But it matters because when your child is not feeling safe, she will struggle to listen. Do remember that you will be morphing in and out of these states on a moment-by-moment basis but that once your SNS has been firmly activated, say in anger, it will take you a number of minutes to return to a state of calm. In this time, it is best to speak as little as possible. If you need to move away from your child to prevent your anger from escalating, do so, but explain it to her so she understands you are not rejecting her but are trying to be kind and stop yourself from hurting her feelings.

How often does your communication have a 'no' versus a 'yes' undertone?

Think of how often the things you say have an underlying flavour of 'no' versus 'yes'. Anything that contradicts, denies, corrects, judges, misunderstands, and generally doesn't resonate with what your child is saying, has a 'no' connotation. Reciprocity, on the other hand, is about a 'yes' undertone. This doesn't mean you agree with everything they say, but that you show them that you recognise and understand it and in doing so, you give them the right to feel their feelings without shame or fear. Here is a small example:

Child, in an excited tone: 'I ran around the playground for the whole lunch-time break today.'

You, sounding vexed: 'When did you eat your lunch then?' ('no' undertone; the implicit message is judgemental – your running isn't very exciting to me and you probably didn't eat all your lunch)

Child, in an excited tone: 'I ran around the playground for the whole lunch-time break today.'

You, light-heartedly: 'That must have been fun! I hope your lunch powered you through all that running.' ('yes' undertone, resonating with the child's emotional state of joy)

Please note you really don't have to repeat back or validate every statement. All I'm saying is that it helps to pay attention to how much of your communication with your child is truly reciprocal and try to engage in connected, non-judgemental dialogue at least some of the time.

CONNECTION EMERGES WHEN YOU SLOW DOWN

I find the word SLOW works well as a memory tool to put this into practice.

S = Soften. Slow down your heart rate and soften your face and voice so that your social engagement system can switch on. Slow your breathing so your heart rate can come down accordingly. Pause a little between breaths and experience a sense of stillness if you can. Depending on your level of arousal, self-awareness and vagal tone, this may be easier or harder for some. But you can control your breath so start with that. When you hug or cuddle your child, try to pause your breathing, soften the tension in your muscles and be still for a few moments.

L = Look and Listen. Once you have slowed your breathing and your heart rate is slower, you will feel a little more relaxed and able to make eye contact. Look at your child's face gently. What do you see in her eyes? Ask open questions and listen to what your child is saying without allowing your mind to make hasty and emotionally charged assumptions that fuel judgement and negativity. Speak less and tune in more to your child's emotional state from looking into her eyes.

O = Open. Open your mind and heart so you can feel connected even in the midst of a difficult situation with your child. The next time she is distressed, try to soften and relax your face and voice, and speak gently, in a lower tone and without the note of sharpness that can so easily creep into the conversation. Adopt a mindset of curiosity not judgement.

W = Warm. Let yourself feel compassion and warmth for your child. She needs you to be regulated and on her side in order to feel safe. Don't judge or shame her because when she doesn't feel you understand her, she is at her most vulnerable (unless she has already hardened herself to you in which case you need to begin very slowly to repair and re-establish connection). When a child does not feel emotionally safe in your presence (via connection), she can't listen, learn and process what you are saying, which renders correction almost useless at that point in time.

Regardless of how old they might be, use your social engagement system to connect and calm your children before you start to correct, analyse, judge or direct them. There is a place for those things, of course, but after connection. You need to be moved by your child's distress but not so much that you are triggered by it yourself. You do not need to feel your child's emotions to the level your child feels them to coregulate your child. You can feel surprisingly light, warm and positive, even in the midst of your child's distress. You can experience pleasure from comforting and soothing your child rather than simply solving the underlying problem.

THE SOCIAL ENGAGEMENT SYSTEM INFLUENCES HEALTH AND WELLBEING

One of the measures of how well the system works for us is our 'vagal tone'. Your vagal tone measures how flexibly your PNS functions and how well it can calm down an accelerated heart rate so that you can return to your equilibrium. Good vagal tone is indicated when you have a small increase in heart rate on the inhale and a decline in heart rate on the exhale, but not too much of either. When there is too much of either, you can slip outside your window of tolerance into either hyperarousal or hypoarousal. Do you see why I've been talking so much about balance? It even comes into play at this microbiological level to regulate several of our bodily, emotional and mental states. When your social engagement system is working effectively, and you have good vagal tone, you will release appropriate levels of oxytocin that support connection with your children and others.

Studies show us that oxytocin plays an integral part in parenting and attunement; when levels are raised, we pay more attention to people's eyes, their smiles, and we view them as higher in attractiveness and trustworthiness. It also calms the amygdala and makes it less attuned to negative threats and more open to positive information[45]. This creates the space for us to feel safe. Oxytocin increases our ability to tune in to human emotions and to recognise and respond better to positive emotional expressions in others[46, 47]. But it also enables us to be more defensive when we are not feeling safe. What makes it so interesting is that it helps us to respond more positively to those with whom we have good relationships whilst increasing our ability to be wary of those with whom we don't.

Warm, connected parenting helps our children to respond better to oxytocin, which in turn allows them to feel pleasure and joy when they are interacting with you and with others. Pregnant women, through increases in circulating levels of certain hormones, become primed during pregnancy to respond to their children in ways that facilitate nurturing through oxytocin release, for example during breastfeeding – changes that are observable in certain parts of their brains, such as the hippocampus. This process is negatively affected by stress during pregnancy and not every new mother is able to release and respond to oxytocin during her interactions with her baby in the way described above.

When parents interact with their children in a synchronous way, with mutual levels of positive engagement and emotional attunement, which can only happen when the social engagement system is turned on, their oxytocin levels also synchronise with each other. The more oxytocin the children are exposed to through loving interaction, the more receptors they build in their brains for oxytocin, meaning they are building the capacity to sustain caring and rewarding relationships with others in their lives. In addition, there is a growing body of research that shows us that oxytocin helps us better regulate our stress response, including improved cardiac regulation by the parasympathetic system[48, 49]. Oxytocin has other wide-ranging benefits. It can protect us against cardiac dysfunction, and it has antioxidant and anti-inflammatory properties across the life span[50].

Your vagal tone predicts wellbeing in more ways than one

Barbara Fredrickson and her team[51] have studied how having higher vagal tone increases positivity and connection, which in turn increases vagal tone. Having higher vagal tone is related not just to better emotional regulation, but because you are able to adapt your physiological responses to fit the challenges you encounter better, it gives you the gift of resilience in the face of stress. A plethora of studies have established that experiencing negative emotions can increase the risk of heart disease and compromise the immune system[52] but having higher vagal tone may positively influence your immune system, including lowering your level of chronic inflammation, which is linked to a host of physical ailments such as heart disease, diabetes and even certain cancers[53]. When your upper vagal system is working well, it counters the health-damaging effects of excessive SNS activation so it is key to managing stress.

On reading this section, you might start to worry that you have lower vagal tone and are destined to a lifetime without the myriad benefits that higher vagal tone can confer. But the good news is that you can increase your experience of positive emotions through the practice of loving-kindness meditation[54] and practising loving-kindness meditation can also improve your vagal tone. This may sound like something you might naturally shy away from, or even sneer at, but it is hard to argue with the science. It is precisely because the same underlying system that calms and restores the body also enables bonding and connection that love and kindness have such a healing and protective effect on human beings. Practising mindfulness meditation and yoga can also help you to cultivate a calm presence, from which your capacity for heartfelt connection can emerge.

DEVELOPING THE ABILITY FOR LOVING-KINDNESS

Certain types of meditation and breathing exercises can bring about a real change in the way you connect with your child, in a sustainable and transformational way, and in a way that reading, by itself, cannot achieve. Emotional regulation and connection involve the body so thinking your way to these states without learning to activate your social engagement system and to regulate your emotional reactions at a physiological level will not yield a demonstrable change in the long term. To really see a difference, try to do a form of compassion meditation at least twice a week or more, especially a loving-kindness practice like the one described here.

I once did a loving-kindness meditation every day for two weeks and found that it brought out the best in me in every sense – as a parent, a partner, a psychologist and a human being. I felt infused to my core with so much compassion and goodwill towards people that I had to work to keep it regulated! For those of us who may not have developed enough receptors for oxytocin in the brain, or who are stressed or depressed, or who struggle with an overactive amygdala, or an inactive insula and orbitofrontal cortex, this exercise may be challenging and will take commitment and regular practice before any benefits might emerge. But I hope you see that this whole book is my attempt to convince you it is worth a try and that you will give it your best shot.

Loving-kindness meditation practice

Sitting in a relaxed position, take a few moments just to be. Inhale deeply and soften your body with each exhale. Start to direct your exhalation into the area around your heart and tune in to what you feel there. Try to soften and relax all the small muscles of your face as you exhale. Let your heart rate and breathing slow down so that your exhale is now longer than the inhale.

Now think of a person for whom you find it easy to feel love and warmth. Bring an image of this person's smiling face into your mind. Notice how this makes you feel. Keeping an image of this person's face and eyes in your mind, **silently** wish this person well by reciting the following phrases. As you recite them, do it with a slow, kind inner voice. Pause between each statement and notice how you feel around your heart. Try to direct all the warmth you are capable of feeling towards this person.

May you be happy.

May you be loved.

May you be accepted as you are.

May you feel safe.

May you be healthy.

May you live with joy and ease.

Notice how you feel around your heart. Perhaps you feel a sense of joy, warmth or contentment. Perhaps you don't feel anything and found the phrases uncomfortable. Maybe you felt overwhelmed, especially if you thought of your child. There is no right way to feel. Just let yourself relax and try to stay regulated. Now repeat the exercise with another person. To cultivate compassion towards your children, try doing the exercise with one of your children in mind.

Compassion for yourself as a child

When you feel comfortable with the above exercise, bring an image of yourself as a child into your mind. Imagine your grown-up self hugging the child you once were and extend the above heartfelt wishes of kindness towards yourself as you were then. Try to allow yourself to relax and feel a sense of compassion for yourself as a child without wanting to shut it down. Some people feel moved to tears by this part of the exercise and that is perfectly normal. It is often rare for us to direct good wishes and compassion towards ourselves so it might feel uncomfortable at first. You may feel moved by this or you may struggle with it; whatever you feel, it's all okay. Remember you know how to regulate your nervous system now so do what you need to bring yourself back to your window of tolerance.

Variation on the exercise: feeling compassion in the face of distress

Once you have practised the above for a few days, you might like to try feeling compassion in the midst of your child's distress, and at a later stage, even a child's difficult behaviour. To do this, follow the instructions as above but instead of imagining a happy, smiling child, imagine your child feeling hurt or upset. Try to regulate what you feel and focus on generating feelings of positivity and warmth, rather than becoming triggered by your child's distressed feelings. You might also like to repeat this exercise with an image of yourself feeling distressed, vulnerable or sad as a child.

Chapter Eleven: Key points

- Play and joyful social behaviour act as a workout for the social engagement system because we learn how to tolerate increasing levels of activation whilst also remaining calm yet alert overall.
- Children can easily move from a state of play into a state of fight or flight because their social engagement systems are still maturing.
- Still, deep states of intimacy and connection are enabled by the upper and lower vagal systems in the absence of threat.
- Pay attention to how your child makes eye contact, and whether their eyes can convey their inner emotional life or are generally quite tense or flat in expression. Notice how they react to changes in your tone of voice and when they become dysregulated. Take action to reduce sources of stimulation, increase sleep and 'non-doing' time, or to work on your own social engagement system so that you can better coregulate them and build their capacity for emotional regulation.
- Children are very sensitive to changes in eye contact that reveal the emotional state of another person. When you are busy, stressed or critical, it will impair your ability to make effective eye contact, without which your children cannot neurocept safety and regulate their emotional state accordingly.
- To build your social engagement system for connection, you need to scan your face and body for signs of tension and use long exhalations to slow your heart rate down. Talking less, making relaxed eye contact, and softening your voice can be helpful.
- Connection emerges most readily when we slow down. Use the SLOW method to put this into practice.
- Being able to switch on your social engagement system, and that of your child, is critical to health and wellbeing. Your vagal tone measures how well your upper vagal system functions and has been linked to heart health, inflammation, positivity and wellbeing.
- You can improve your vagal tone through practising loving-kindness meditation and other breathing exercises such as those I've presented in this book.

CHAPTER TWELVE

Drive and strive, threat and defence, calm and connect – what mode are you in?

In this chapter I'm going to present you with a model that ties everything we've covered together and what's more it is easy to remember and to make practical use of when you are with your children. It is based on the wonderful work by psychologist Paul Gilbert on compassion[55] and I have found it so useful in terms of understanding what gets in the way of me staying calm and relaxed around my children. This will help you to think about how you use your time, what matters to you in the way you live your life, and how what you value will shape the extent to which you demonstrate heartfelt parenting.

THE THREE SYSTEMS OF EMOTIONAL REGULATION

There are three broad emotional systems that we move in and out of on a moment-by-moment basis. Each of these systems links to different groups of emotions and they are underpinned by the activation of the sympathetic (SNS) and the parasympathetic (PNS) branches of the nervous system. Each system is linked with different motives, reactions, feelings, thoughts and chemicals. These states are **Drive and Strive**, **Threat and Defence**, and **Calm and Connect**, all three of which evolved to facilitate our survival in different ways.

Drive and Strive System

State of activation.
Goal-focused, seeking,
striving, achieving, doing, consuming.
Emotions/states: enthusiasm,
excitement, pleasure,
anticipation, self-satisfaction,
restlessness, agitation,
compulsion, rigidity

Calm and Connect System

State of safety and relaxation.
Non-striving, non-judgemental,
trusting, present in the
moment, connected.
Emotions/states: love, warmth,
empathy, compassion, bliss,
contentment, openness

State of fight, flight or freeze.
SNS or lower vagus.
Emotions/states: anger, fear,
disgust, contempt, stress,
judgement, closed off,
defensive

Threat and Defence System

Making sense of the three systems:

Do you recall how we explored the difference between Trait (more permanent and stable characteristics in you) and State (transient physiological and mental changes) in Chapter One? That difference applies to these three emotional systems. We move in and out of these three circles in a dynamic, variable way, depending on our genetic predispositions, attachment styles, personality traits and emotional regulation styles (trait) or we might move in and out because of the situations we find ourselves in, or our moods and biological states such as hunger and tiredness (state). Our general state at the time will have an impact on which circle we move into and how well we regulate ourselves in and out of it. When we have developed good emotional regulation circuits, we will be able to shift between these states in a flexible way that is adaptive and constructive to the situation we find ourselves in.

However, we can also get stuck and reactive, particularly in drive and strive or threat and defence modes, which becomes problematic in the context of parenting. The upper vagal circuit determines how we slide in and out of these systems without becoming dysregulated or, worse still, disabled in one mode.

THE DRIVE AND STRIVE SYSTEM

This is essentially the human motivation or approach system that has evolved to ensure we are adequately motivated to seek food, shelter and to reproduce, all essential ingredients for survival. It is goal-focused and is associated with seeking, striving, doing and achieving. This system ensures we are motivated to get out of bed in the morning, get dressed, go to work, take care of our children's needs, seek social interaction, self-esteem, approval, status, and all the other things we strive for in life, whether they are physical or psychological goals. What we strive for is unique to us and depends on our values, personalities and attachment styles.

As this is an action-oriented system, it is underpinned by either the upper vagal system with the vagal brake, or the activation of the SNS that ramps us up for action, movement and grasping. The emotions associated with this system are generally positive and rewarding, e.g. anticipation, excitement, self-satisfaction, self-confidence, optimism, dominance, pleasure and the type of happiness that comes from obtaining material possessions, approval or status. When we are in this mode, we can feel purposeful, strong and energised. This system relies on the neurotransmitter dopamine, which is the chemical we release in anticipation of, or in response to, a rewarding experience. Because seeking and striving, at least for food, is so essential to our survival, dopamine contributes to positive reinforcement and can also be quite addictive.

Driving and striving can make us self-focused and impatient

Driving and striving lateralises more to the left hemisphere than the right, which is not surprising given the left evolved to allow us to pay a narrow, focused attention devoted to grasping, obtaining (e.g. food), mastering and controlling things. People who are highly driven to get things done, either by their nature or in that particular moment, may become more left-dominant in their interactions with people. When we set ourselves goals and become too focused on achieving them, whether they are minor goals such as tidying something up, or significant goals such as being promoted at work, we are liable to tip from drive and strive mode into threat and defence mode when we encounter an obstacle or challenge.

People who become rigid in their desire to do things at a certain time, or in a certain way, can become so preoccupied with their own thoughts and needs that this system becomes dysregulated; people around them become barriers to them achieving what they need to get done, creating frustration and lowering empathy levels. In this mode, other people become like objects to be dealt with; to be persuaded, removed from our paths, or 'managed' in order to achieve what we are striving for in that moment in time. We become resentful and irritated, even angry, because of what we see as interference with achieving our objectives. Something as simple as another person expressing a contradictory opinion, or having a goal that doesn't match ours, can set us off. How regulated we remain in this state depends on our tolerance for high activation states and our innate levels of motivation. We are not all built the same way; some of us need to be very busy, active and goal-focused and others value and need, even at a physiological level, calmer, slower states to feel balanced.

When we become dysregulated and stressed because we have too much to do, the social engagement system is compromised; we are less open to whatever is emerging in the moment and we make poor listeners. We listen to hear what we need to hear, or to quickly appease people, but we struggle to listen with a sense of genuine empathy, openness and acceptance. This mode, at least at a mental level, can generate a need for control, impatience, restlessness and, at the extreme end, rigidity and aggression. I can think of many highly successful executives I have worked with who become so dysregulated in drive and strive mode that although they are excellent at getting things done, they have a tendency to alienate and upset people along the way, something they are often unaware of.

In parenting terms, it is very simple to think about when we might easily become dysregulated in this mode, because it probably occurs in every household every morning when parents need to get their children out of the front door and into school on time. When we have a time constraint coupled with multiple tasks to achieve, we can become mindless and easily frustrated, perhaps shouting at our children to get them to hurry up, or rushing them and telling them what to do with little sense of them as individuals at that point. Some of us are so chronically busy with things to do that we are almost permanently in a dysregulated state of drive and strive, which then lowers the threshold for us to tip into the threat and defence mode. Even having things on our minds, and being interrupted by our children, not physically but mentally, can be a problem.

The rules we make up can dysregulate ourselves and our children

The more rules you set for yourself and your children about what you and they need, must do, should do and cannot do without, the more likely it is that you become dysregulated in drive and strive. Children are not born with a huge inner motivation to have and do all the things we impose on them in their early years. In some countries they are coerced into this mode of 'doing' and 'striving' far too young, when what they need most is to be able just to be, to take in information through their senses, and to connect with and be curious about the world around them. It is this ability to engage fully with the world that develops their brain circuits for regulation and wellbeing in a balanced and integrated way. When we raise children with an excessive emphasis on drive and strive, we are not only creating future problems for them but we are also changing the way humans and societies engage with each other, because it is difficult to sustain a genuine connection in this state, particularly if we become dysregulated by it.

How does our fixation with pleasure and comfort get in the way of wellbeing and connection?

Although this system evolved to motivate us to seek things that enabled survival, and feeling pleasure when we achieved them motivated us to continue seeking them, pleasure itself has now become the goal rather than the by-product of striving for something that is necessary for survival or advancement. This constant cycle of being busy, doing activities, achieving, owning and controlling is damaging because we become self-focused and less willing to tolerate personal discomfort, which changes how we engage with family and others with whom we have close relationships. It is hard to really connect with each other in a compassionate and meaningful way when we are in this mode too much of the time. Because striving stimulates the dopamine and sympathetic nervous system, which ramps us up and makes us want to move faster, do more, have more pleasure, and achieve more, it can have an impact on our stress levels and our heart health. Sometimes the more we have the more we crave because the law of diminishing returns applies here; humans habituate to the level of pleasure they become accustomed to and then need even more of it to create the initial surge of positive emotion they experienced at first.

Being too busy comes with a price to our health, relationships and wellbeing

It seems to have become fashionable to be constantly busy: travelling; taking children places; playing with them; reading to them; ferrying them to and from activities and so on. If we have a strong upper vagal circuit, we may be able to remain regulated and connected regardless. But for so many of us, this is not the case. It is not hard to think of examples of parents who are frazzled, tired and stressed by the daily pressures of looking after children in this way and balancing it with all the other demands we face and place on ourselves, such as working, exercising and socialising. If we are becoming stressed by doing too much, even if we think we are doing these things for the good of our children, there are pervasive and damaging consequences to living life in this way, one of which is compromised connection.

Being under pressure to do too much and to do it all too fast creates stress, which shuts down the social engagement system. This is creating a fundamental imbalance in the functioning of our brains and nervous systems that impacts not only mental wellbeing but also our physical health. I have come across too many people who are stressed at work and begin to experience depression and anxiety because of it. They cannot pause and reflect; they cannot relax and be themselves; they are under pressure to continue to compete, to innovate, to self-improve and prove themselves and there is no let-up. This is not compassionate or wise, because we are not different beings at work than we are at home, and our experiences and emotions in one arena will inevitably seep into the others.

When our children become used to a high level of drive and strive fuelled by being constantly busy and always having something to do, they will struggle to relax and connect with anything, including their inner selves. This is when they 'act up' at home or become addicted to their phones and other gadgets, because they have not learnt how to access their calm and connect system to be able just to be in the moment. They struggle to tolerate boredom because they don't know how to generate positive emotions from within, in the absence of structured or pleasure-inducing activities. It is worrying that our children don't learn how to delay gratification because so many of their experiences provide immediate gratification in a way that our brains are simply not used to. They are certainly not used to waiting for much in life. I walk my children to school every day and on the odd occasion we need to get the bus, one of my children likes to insist we check the 'bus app' to see how long we need to wait before it arrives. Although we have a regular discussion about tolerating discomfort and uncertainty this seems an outmoded and alien concept to my child!

Children need 'downtime' to balance and reset their nervous systems

My view is that children need to spend a lot more time relaxing and resting at home. When they are not accustomed to downtime and cannot regulate themselves when they are bored, it is hard for parents to tolerate this and often they react by taking children out or scheduling activities for them. But if we continue to do this, they don't learn how to calm down, engage their right hemispheres, activate the social engagement system and get on with being in the moment, something we are naturally designed to do and that is fundamental to our wellbeing. Yes, I do know some children manage this better than others but, still, we can encourage it, however tricky it might seem. A lot of the time, children are agitated at home because you are agitated; you are agitated because you are anticipating they will be difficult, have fights, drive you mad and it will all degenerate into chaos. Perhaps you compensate with rigidity in the form of schedules, structure and organised activities because you don't like the feeling of being out of control.

A few years ago, when I first started putting all this science together for my clients, I made a conscious resolution to slow down with my children. I have had to teach myself how to relax – and I mean really relax; not just looking as though I am – in the company of my kids at home, without trying to do anything at all. I have seen how they come to soften and settle down when I am calm and genuinely open, especially when I have nothing on my mind. Even my 'drivey and strivey' child becomes stiller, calmer and slows down a little. This didn't happen overnight and I only attempted it once I had already developed my emotional regulation capabilities. I actively resist too much planning and scheduling, and we find that this spontaneity gives us the freedom to live in a way that respects our emotional and physical states at the time.

Allow your children time to relax without planned, structured activities

Too many scheduled and structured activities encourage our children to ignore their bodies, emotions and needs, and create a left-hemisphere-dominant way of engaging with ourselves and the world. I'm not advocating that we encourage them simply to give in to their feelings all the time either; no, they must learn to tolerate the discomfort of doing things that they might not feel like doing, but they already have a lot of structure imposed on them through the schooling system and I don't believe they need much more of it in their spare time. Instead, they need to learn to tune in to their instincts, values and emotions and to listen to and respect their bodies. Creativity emerges when they are in the moment and boredom can make them more appreciative of small things that they might otherwise find dull or boring in a busy, fast-paced life. Spending time in green, open space is also soothing and helps lower stress levels.

I do understand that some children bicker and fight and struggle just to be in the moment. I see that some children have a need to move around and keep up a level of SNS stimulation to feel okay. I don't like to be prescriptive because it only leads to more striving and stress but do be aware that your internal state influences your children's capacity for regulation at home. Regardless of their personality traits and other influences, they can be shaped by the quality of emotional regulation that you demonstrate around them. When you are genuinely relaxed at home, they will take their cue from you. If you overreact every time they bicker or become dysregulated, or you take them out all the time, they internalise the message that having downtime is boring, stressful or bad, and this will prime them towards emotional dysregulation around you at home in the future. As parents, we must help our children regulate their states but must be careful not just to give in to them at the first sign of frustration. Striving and achieving is important but it must be balanced with another system that we will come to later, the calm and connect system.

THE THREAT AND DEFENCE SYSTEM

As we've been through in Chapter Ten and Chapter Eleven, the threat and defence system involves the fight, flight or freeze response. Emotional states associated with the threat and defence system are anger, rage, hostility, aggression, stress, anxiety, fear, emotional avoidance, depression, hopelessness, shame, despair and dissociation. When the fight or flight response is activated, the stress system is also switched on, resulting in hormonal changes, for example increases in cortisol that, if persistent, can have a significant impact on health and wellbeing. Cortisol suppresses the immune system and downgrades digestion and reproduction; it also sends signals to the brain to heighten the activity of the amygdala leaving us even more alert to potential sources of threat[56]. Stress, because it leads to a reduction in upper and lower vagal activity, leads to sleep disruption, which in turn has a cascading effect on mood, appetite, digestion and inflammation. Because stress sensitises the amygdala even further, it is not hard to become stuck in a cycle of low mood, negativity, self-criticism and impaired connection.

Sources of stress can be self-generated

Although the dangers we faced during our early evolution were mostly threats to our lives, these days, at least in the west, the threats and sources of stress we encounter are typically psychosocial in nature. Our triggers in this day and age are threats to our self-esteem/egos, to our need for control, for certainty and for status (the house, the grades, the job, money). Because these ancient brain systems evolved in times of great scarcity, we are sensitive to where we are on the status ladder, whether we are being treated fairly, and whether we are accepted as part of a social group, because all these things increased our chances of having a greater share of the limited, life-saving resources we had available. We also know that our early days as humans were fraught with aggression and the instinct to fight is linked to the instinct to protect ourselves. But these are days of abundance, control and comfort and many of our sources of stress are generated by our own minds.

We tip into threat mode when the amygdala signals we are not safe. Feeling safe around your children is vital to good parenting because when you are in threat mode, your children sense it and mirror that state back in some way. So why do we respond to our children in a defensive or threat-based way? In the context of the stressors we have faced over the ages, there is nothing objectively stressful about bringing a child home from school, cooking a meal and putting her to bed. It is our own internal demands and reactions to these things that lead to the perception that they are threats. It is because we have thoughts and judgements about what will or might happen next, or how what is happening shouldn't be happening, that things become stressful. It is the internal mental demand that things should be easy, comfortable and predictable that can make being around our children a stressful experience. It is that we sometimes have priorities other than our children that makes caring for them difficult. And of course, we can all catapult into a dysregulated state from being triggered by the emotional states of our children, which as we now know will relate to our attachment styles and general patterns of regulating emotion.

Sometimes parenting becomes difficult because we have certain expectations of our children and we tip into a threat state when those expectations are not met. There are too many pressures for children to be perfect and do things the right way: they must be kind and agreeable; they must be self-confident; they should look right (this is a source of mental pressure for children these days because of the left-brain way they relate to their bodies); they must achieve and stretch themselves; they must not bicker and fight; they must listen and learn; they must eat a balanced diet; they must exercise and do a range of activities; they must read books from the age of four; they must go to school, get good grades and carve out a solid future for themselves (we often expect them to follow a rigid, linear pattern of achievement that puts immense pressure on them), have a good social circle and so on. The more we expect from them, and they expect from themselves, the more likely we and they are to go into threat mode when things don't seem to be going to plan. This generates anxiety, worry, negativity and stress. And the more we go into threat mode around them, the less likely we are to have a peaceful, open, loving relationship with them. And I hope I've said enough by now about what this means and why it matters.

THINKING ERRORS THAT CONTRIBUTE TO STRESS

We know that managing emotion requires bottom-up soothing at a bodily level. But once you've done that, it can be very empowering to learn how to sharpen the input you receive from your top-down system, driven by your prefrontal cortex, the brain region responsible for rational thought, balanced thinking, self-control and modifying our thought patterns. Here's an insightful look at some of the habitual thinking errors we make and how, in changing how we perceive things, they tip us into threat and defence mode, often in an irrational way that is not grounded in evidence or logic.

These errors are based on mental shortcuts that can sometimes help with information processing but which can also lead to unnecessary anxiety, stress, anger and dysregulated emotion. These thinking habits occur outside our conscious awareness but with some practice you can become aware of when you are making them and how they alter the way you perceive things. Do note children make these sorts of errors frequently because they lack the mental flexibility and sophistication to challenge their initial assessments of things. Do also bear in mind the role of **confirmation bias** in generating and maintaining several of the thinking errors below. Confirmation bias is a fundamental human error in reasoning that leads to unwittingly selecting, seeking, explaining, recalling and memorising information in a way that fits with what we already know, what we expect or what we want to see.

Polarised Thinking: This is also called 'black or white thinking' and is based on dichotomous choices, i.e. all or nothing, good or bad. We tend to perceive things at the extremes, with very little room for a middle ground. This makes us judgemental and uncompromising in our assessments of ourselves and others around us. Examples might be, 'If I'm not successful I'm a failure' or 'If it is not done in this particular way, it is useless.' When it comes to our children, many of our anxieties about them will involve such dichotomous thinking. If we accept them as complex fallible human beings with positive and negative qualities, this relieves some of the pressure to raise perfect kids who must achieve positive outcomes at all stages of life.

Overgeneralisation: This is when we come to an overall general conclusion based on a single incident. If something negative happens once, we begin to worry because we see it as a generalisable pattern and therefore expect it to happen again. Thoughts such as, 'Nothing seems to go right for me at the moment' are an example of overgeneralising, as is using words such as 'always' and 'never'. When it comes to our children, if we feel they are always behaving badly or doing things well we are likely to be overgeneralising. We must learn and teach our children that things happen in degrees and shades of grey. They are neither wholly one thing nor another. They must be allowed to make mistakes, be fallible and get things wrong without thinking that means something broader about them and their chances in life.

Mind Reading: We assume we know what people are feeling and why they act the way they do even though they have not said so and there is no tangible evidence to support our assumptions. This is often fuelled by our own insecurities and mental models that lead us to expect certain things, and to interpret what occurs in light of our existing explanations. Mind reading depends on a process called projection. We imagine that people feel and react the same way that we do. Mind readers arrive at conclusions that ring true for them, without checking whether they are true for the other person. Examples of this might be, 'I know they're just being difficult on purpose' rather than thinking, 'Maybe they don't like doing this and I need to ask why.'

Being Right: This happens when we feel we must prove that our opinions and actions are correct. Having to be 'right' often makes us poor listeners. When we're trying to prove ourselves, we aren't interested in the possible truth of a differing opinion, only in defending our own. Being right becomes more important than building relationships. Children frequently struggle with the notion of being wrong and can become argumentative and belligerent when we contradict them. This is a sign of an immature brain and we mustn't get caught up in battles such as these with our children. It is far better to ask them a question that gently challenges their thinking than to contradict them harshly.

Demanding: This is when we have a list of internal unconscious rules about how we and other people 'should' act. We feel annoyed when people break these rules, and we feel guilty if we violate our own rules. Cue words indicating the presence of this distortion are 'should', 'ought', and 'must'. To counter this type of rigidity, we must remember that there are no laws of nature or the universe that dictate things should be a certain way because we would like them to be. This thinking error contributes to most emotions on the anger scale – annoyance, irritation, frustration and rage. An example of an underlying thought that stems from demanding is, 'People should always do the right thing.' I find it helpful to imagine that not long ago in our evolution, we were clubbing each other on the head for a piece of meat and yet we now find ourselves here demanding fairness in all interactions and becoming enraged over simple transgressions that are really not anything other than an inconvenience. These rules may have evolved to maintain fairness amongst humans but imposing them with rigidity rather than compassion only leads to more anger.

Catastrophising: We expect negative things to happen and at the first sign of things going wrong, we blow them out of perspective. Feeling anxiety is linked to predicting negative outcomes without sufficient evidence. When we do not trust in our capacity to cope with difficult or uncertain situations, the negative emotions we might experience, or our ability to adapt to change, we might find ourselves thinking the worst is about to happen. An example of this type of thinking might be, 'This is going to be a disaster – what if my child fails?'

Blaming: This happens when we hold other people responsible for our difficulties, or take the other tack and blame ourselves for every problem. Blaming often involves making someone else responsible for choices and decisions that are actually our own responsibility. Blame can be one way of protecting our egos. An example might be, 'This would never have happened if he just did what I told him to do.' I have come across parents who are constantly on the hunt for someone to blame for their children's lapses and failings because they can't tolerate the notion that they as parents, or their child, may be fallible. They expect that everything should be done perfectly by the people and institutions around their child so their child can avoid the experience of emotional or physical discomfort. Whilst this is understandable, it is not helpful to the development of resilience.

Filtering: We selectively pay attention to the negative details and amplify them, while filtering out all positive aspects of a situation. We might become preoccupied with one single detail, and in isolating it from the overall context, give it more significance than it might otherwise warrant.

Personalising: This is the tendency to relate things around us to ourselves even when this is not the case. This often involves thinking that the things someone else says or does are a reaction to us or some indication of whether people like us or not. For example, 'She didn't say hello to me this morning but she was chatting happily with someone else – probably doesn't like me as much. Maybe I'm not that popular and likeable.' Bear in mind these are not conscious thoughts, but you can begin to identify them from the way you feel and react to things. Children are prone to a high level of personalising because their prefrontal cortices are still in development and they are not able to generate alternative perspectives. When you are dysregulated, they will tend to assume that it is in reaction to them. What this means is that you need to provide them with alternative explanations where that is reasonable and truthful.

(Adapted from McKay, Davis & Fanning 2007 and David Burns 1999)[57]

Questions to challenge negative thoughts and assumptions

If you think back to the stages involved in emotional regulation – noticing and accepting, soothing and calming, challenging thinking patterns – this section helps you with the third stage of refining and challenging your own thoughts, interpretations and beliefs, which will provide a strong, top-down mechanism for managing states of threat and defence. Do recall, though, that if you are in a heightened state of threat, your higher thinking capabilities are temporarily suppressed so you must first calm the body back into your window of tolerance before trying to manage your thoughts. Once you are beginning to calm down, take the time to coach yourself into thinking in a more balanced, evidence-based manner. Then use this ability to coach your children over time. It is useful for them to learn to challenge their own assumptions and evaluations, not least because irrational, negative thinking patterns are a hallmark of mental health issues such as depression and anxiety. Here is just a small selection of questions that can help elevate us out of threat mode:

- What real evidence do I have to support my view of things?
- What evidence can I think of that does not support my view?
- What other perspectives might there be on this? How might someone else see it differently?
- What other explanations might there be for this? Am I automatically assuming my interpretation must be the only correct one?
- Am I distorting information to fit with what I want to believe? Is this an example of confirmation bias?
- Is my ability to think rationally being hindered by my emotional needs/fears/insecurities? How would I see this if it were happening to someone else's child?
- How likely is it that my predictions will actually happen? Where is the evidence to support this?
- What is the worst thing that will happen if my prediction does come true?
- What is the most likely thing to happen?
- Am I looking at positives and negatives in balance – or I am focusing more on negatives?
- Am I worrying about things that are within my ability to influence?
- Will this thought help or hinder me or my child in achieving my goals?
- What advice would I give someone else in this situation?
- Am I fretting about how things should be rather than accepting and dealing with them the way they are?
- Am I judging myself, or my kids, or the situation, as a whole based on only one example or instance of something?

- Am I being open-minded and willing to consider other views and possibilities? If not, what am I holding on to?

(Adapted from Fennell, 1989)[58]

THE CALM AND CONNECT SYSTEM

We've talked about the drive and strive system and how it motivates us to achieve, grasp and strive. We've also talked about the threat and defence system and how it has evolved to protect us from threats to our lives. Both these systems are stimulating systems, based on putting a foot down on the accelerator and ramping us up for action. But what about the brake? When do we allow ourselves to calm down, rest, relax and connect and how much of a balance do we get between this way of being and the other two states? Some of you might be surprised to think about this third state because it seems we spend so much time in the two modes already discussed that there would be little time left for much else. But this is not what we are made for and we must take the time to think about how we bring back some balance to our lives, because it is vital that we nurture this system that is so essential for wellbeing, resilience and connection. When we become too caught up in drive and strive mode, or when we are reacting from a place of threat, our calm and connect system, which is fundamentally a right-hemisphere way of being, is suppressed. This is when parenting will feel tough and the children will begin to become dysregulated and oppositional.

The calm and connect system generates a sense of contentment, stillness and openness without any judgement, pressure, or sense of striving. It is about a right-brain sense of presence in the current moment with an acceptance of things just as they are. It is a state of 'being' rather than 'doing' and it is activated when your upper vagal system is in the lead. This soothing/affiliation system, as Paul Gilbert calls it, is rooted in the feeling of being lovingly coregulated and cared for, either by yourself or by others. It is this feeling of being shown loving-kindness that accounts for why this state is so pivotal to good emotional regulation; we cannot return to our window of tolerance without our bodies knowing, mostly unconsciously, how to calm ourselves back to a state of equilibrium.

Parenting rests mostly upon this calm and connect system because when we are still, we can notice and tune in to what our children need and feel. But more importantly it is from this state that we can feel moved by the experiences of others around us and want to alleviate them, which is the basis for compassion. Although compassion involves a sense of striving to alleviate the distress of others, it is an integrated state and may involve balanced activation of the left and right hemispheres, enabled by the social engagement system. Drive and strive, and threat and protect modes are mostly self-focused. Although you can be sociable and playful in drive and strive mode, and you might find this rewarding, it is primarily about pleasure. Finding social interaction rewarding and pleasurable is important, especially so in parenting, because our children need to feel that we enjoy being with them and find their company pleasurable. But the calm and connect system, in my opinion, is even more important for the restorative health of our children's bodies, brains and minds.

The calm and connect system facilitates emotional synchrony

The difference between the drive and strive pleasure-based social connection and the calm and connect mode is that the calm and connect system allows you to be moved by things rather than simply feeling pleasure from them. When you are moved by something, you feel in synch with it and transform from a sense of 'I' to 'We'. In such moments, you and another person can experience something as 'one' because you come to be matched not just in your joint attention towards the experience but also in your inner physiological response towards it. Paradoxically, the state of experiencing things as a 'We' strengthens your child's sense of 'I'. The calm and connect response is facilitated by several chemicals and hormones such as acetylcholine, GABA, which has an anti-excitatory, calming influence on the brain, and also oxytocin and vasopressin.

HOW DO YOU MOVE IN AND OUT OF THE THREE MODES?

Naturally we can't always be in the calm and connect mode and neither do we want to, because all three systems play their part in ensuring our survival. But our lives today are filled with sources of stimulation and our minds generate ever-increasing demands so that over time we as humans, and even whole societies, are moving away from connection and caring, towards selfishness and disconnection. Many of us have a level of dialled-up amygdala activity that means we don't know what feelings of loving-kindness or compassion feel like, and we don't demonstrate it enough towards ourselves. When we can't demonstrate it towards ourselves, I question how well we can truly feel it for our children, at least in a consistent and regulated way. What we need for good enough parenting is to spend most of our time with our children in a combination of drive and strive and calm and connect, with the flexibility to switch in and out of them moment to moment. This rests on your social engagement system. Threat must feature only when there is good reason to feel angry, anxious or disgusted and even then, we must stay regulated because children are quick to feel anxiety and look to us to calm this in them, not to escalate it.

What does it look like when we move in and out of the three states with our children? Just today I have seen how having too much to do impairs connection with my children. Writing this chapter has been a race against the clock; I was so focused on getting it finished this evening that I tipped from drive and strive into threat and defence mode with my child. He was arguing with me over something quite insignificant in the wider scheme of things, but rather than putting it in perspective and letting it go, I overreacted and ending up telling him off. I could see, on some level, that he was getting upset and I could certainly hear the threat in his voice because it was raised. But I didn't stop in time and he went into hyperarousal.

On reflection I can see that he became a threat to me in that moment because I had too much on my mind and I wanted him to stop talking and go to bed so I could carry on writing. I work hard to keep myself regulated but when I have too much to do, and I have combined that with other drive and strive activities that stimulate the sympathetic nervous system (intense exercise is an example of this), things build up and staying calm with the children can become more of a challenge. Given we are in an almost unceasing cycle of SNS stimulation, it is becoming harder for so many of us to come back to our inner baseline.

This is where we must remember the natural cycle of rupture and repair that characterises all human relationships. My shouting created a rupture in both mine and my child's calm and connect system that I knew I had to repair. Once I had calmed down (I had to do my breathing exercises for a few minutes to make sure I had completely softened and relaxed), I went to see him in his bedroom. In a regulated state, and ensuring the repair process was focused on **his** feelings, and not my need to feel better about myself, I explained how sorry I was and empathised with how awful it must have felt to him to be shouted at by me, someone he looks to for safety. I made sure my voice was gentle and slow and I was looking at his face to gauge how he was feeling.

I asked him how he felt and he told me it hurt him physically in his ears, which makes sense, when you think about the social engagement system involving the muscles of the middle ear. I listened to him and gently stroked his ears, and explained this was not his fault and he didn't deserve to be talked to like that. I explained that I was under a bit of pressure. Once he was calm, I did also point out that he needs to think about how repeatedly arguing back with a grown-up who has possession of the facts in a way that he may not is not a sensible strategy. We talked about it for a minute or two, after which he suddenly changed the subject. This was my cue that he was feeling okay and the time for repair was over (I would have preferred to go on about it a bit longer but this was about him, not me). When I left his room, he was quiet and calm and I felt comfortable that most of the anger and shame he had felt during the episode had ebbed away. We were back in calm and connect mode.

Do you know which of the three modes you are in with your children?

Take a moment to think about which of the three modes you spend most of your time in. Reflect on how often you and your children have opportunities just to 'be', rather than doing, thinking, rushing, planning or worrying.

- When you are with your children, how often do you feel you are busy and preoccupied or going through the motions?
- How much of your time do you spend telling them what to do, hurrying them on or giving instructions?
- When you are in drive and strive mode, what changes in your behaviour and body language? Does your body become tense or rigid? Does your voice change?
- What happens to your ability to listen with an open mind?
- When do you find yourself becoming most frustrated and tense around your children?
- When do you feel angry, anxious or stressed by your children?
- How often are you able to relax completely around them, with nothing on your mind, and nothing to do except connect with them?
- When do you feel most connected and calm around them?
- What are the barriers to you being in the 'calm and connect' mode with them?
- When are your children most likely to enter the 'calm and connect' mode?
- What can you do to get a better balance between the three systems?

Chapter Twelve: Key points

- There are three broad emotion systems that we move in and out of, each representing a different pattern of emotions and underpinned by a different pattern of nervous-system activation. These three systems are: drive and strive; threat and defence; calm and connect. We move fluidly in and out of these states depending on our traits and also the states we find ourselves in at the time.

- The drive and strive system is the human motivation system. It is associated with seeking, striving, achieving, pleasure, control and anything that is goal-focused. Anything that you think you need, want to do, should do or should have will activate the drive and strive system. Children can get caught up in this drive and strive culture of achieving, acquiring, doing, competing and being busy, which means they are not getting the time they need to develop a balanced, integrated brain based on the right-hemisphere state of open presence and just 'being'. Try to reduce planned, structured activities for you and your children to allow them more time to relax at home.
- The threat and defence mode activates the fight, flight or freeze mechanism. Much of the time this occurs because we generate rules about how we want and need ourselves, our children, and the world around us, to be. This creates pressure on our children.
- Our thinking patterns can generate, maintain or exacerbate states of threat and stress. Understanding the thinking errors you make, and being able to challenge them using reflection and reasoning, can help you to regulate emotions more effectively.
- The calm and connect mode is about 'being' rather than 'doing' and is underpinned by the social engagement system. It is from this state that we are capable of empathy, compassion, soothing and connecting.
- Parenting requires a balance between all three states, with an emphasis on the calm and connect mode and reasonable levels of drive and strive. To be able to inhabit this state with our children we must be able to feel it within, and towards, ourselves.

CHAPTER THIRTEEN

Creating the conditions for heartfelt connection to thrive

Coming to this last chapter, I feel as though we've been on a journey together. I don't know about you, but the process of writing this book has made me reflect considerably on my own parenting style. Some of it has been tough to acknowledge, but I'm not in this to judge anyone, including myself. I started exploring my parenting style to learn and change so I can get closer to providing my children with the emotional safety and connection I know they deserve to experience. What keeps me motivated is knowing this is achievable and these methods I've presented in this book work. Most of us are able to seek and find ways to love and nurture our children, even though our brain-body-minds want to push and pull us in different directions sometimes. We need to be strong when they need boundaries, challenging when they need to push themselves, and gentle when they are feeling vulnerable. But most of all, we need to be tuned in so we can notice what they need in order to feel safe and loved and extend this to them when we are able.

When we stop actively cultivating a state of calm and connect, life will find a way to arm-wrestle us back to dysregulation all too easily. Family life might become more difficult, tiring and frustrating. Because we now live in ways that are so oriented towards the drive and strive mode,

231

and our worlds are now organised to promote this, learning to relax and calm ourselves must become a habit that we devote time to, just as we might do with exercise or personal hygiene. Think of it as nutrition for your emotional brain. If you work on your capacity for emotional regulation, compassion and openness, you will become so much better at noticing early signs of dysregulation, stress or disconnection and repairing them through managing your emotional state. What I'm trying to say is that maintaining the three conditions for heartfelt connection: openness, emotional regulation and emotional safety, needs effort and commitment. And just as regular practice leads to new neural circuits via neuroplasticity, when you don't use it, you lose it, so you can find yourself going backwards.

In writing this book, I wanted you to feel moved by how deeply wired we are for connection and how integral it is to our health and wellbeing. Whether or not we can shape our children's brains through connection, I hope you've come to believe, as I do, that connection is a worthy enough aim in itself because it is at the root of emotional and psychological safety. I hope that you will commit to making some small changes in your life to allow yourself and your children to benefit from what the neuroscience reveals is such a pivotal part of what we need to live a good life. I am convinced that our busy left-brain-generated goals and ways of viewing the world are undermining the close connections we need to stay 'human' in the way we currently know it. We need to focus on connection even more in an age where speed is king and our appetite for comfort, control, pleasure and certainty conspires against our desire to nurture our children for resilience and wellbeing.

Let's not forget why we started down this route to begin with; we wanted to connect with our children and in doing so, to empower them with a lifelong resilience in the face of stress and pressure. We now know that not only will heartfelt connection build their brains for emotional resilience but it makes them and us happier, healthier, calmer and more accepting of ourselves. So, what do we have to do to ensure we don't lose the gains we've made so far? What will you need to do to maintain a loving connection with your children in the face of all the pressures and distractions, your own insecurities and childhood history, and those characteristics and traits in your child that seem to clash with your own?

In this final chapter, I'm going to be a bit more prescriptive and outline ways in which we can adapt our lives to enable heartfelt parenting to emerge naturally. Some of these methods require more effort than others, and all of them require at least a modicum of dedication and commitment. But what they all have in common is that they are designed to benefit you first. You can only extend heartfelt warmth, love and kindness in a regulated way to your children when you are topped up with the ingredients you need in order to thrive and flourish. These include elements that affect your brain, your mind and your body because they are all interconnected and will influence each other when they are out of kilter. I think you know by now that your own wellbeing, and in particular your stress levels, have a significant impact on your ability to regulate your emotions.

NURTURING YOURSELF FOR CONNECTED PARENTING

Actively and regularly practise emotional regulation by calming your nervous system

All this reading and thinking will not leave a lasting impact on you or your children without you addressing parenting at a physiological level. Connection is a bodily thing and for it to work you need to learn to regulate your emotions, calm your nervous system and allow what we humans are designed to do to emerge naturally, all of which requires bodily awareness, not just intellectual insight. It has to be led by your right hemisphere rather than your left so you can't fake it, you can't force it and you can't read your way to it. Believe me, I've faked it so spectacularly that I have even fooled myself but not my children; they always responded to what was going on inside.

If you wanted to get fit, you wouldn't simply read a book about it, you'd have to get up and move and you'd have to do it repeatedly. It's much the same with building emotional health. So please take a few minutes, every day if you can, to learn and practise emotional regulation. I have shown you how to reset your system and strengthen the brain circuits that underpin connection and you know it involves calming and balancing your nervous system using your breath. If you're starting to feel irritated at having to do something tangible, I fully understand. You already have enough on your plate and making time to do breathing exercises might make you feel impatient and restless, especially if you're chronically wired up for action and being busy! But there's concrete proof that this works and I hope by now that you are convinced that this approach is based on sound science.

Do whatever you can to cultivate a mindful, open presence so you can better tune in to your emotions on a day-to-day basis. This is not an easy task in the fast and demanding world we live in today because almost everything you do will be activating your sympathetic nervous system, which lowers your threshold for stress and for the activation of a fight or flight response. All the breathing exercises I have presented you with in this book are on my SoundCloud and I recommend you listen to them until you are very familiar with regulating yourself independently.

You might also like to download an app that guides you through various mindfulness or emotional regulation exercises. You don't always have to set time aside for this. Try to incorporate it into your daily life by softening and slowing your breath, checking in with yourself and staying present in the moment. Personally, I've noticed that every time I stop practising some form of meditation or brain-balancing breathing exercise on a regular basis, however well I feel initially, it inevitably affects my relationships and my inner sense of wellbeing a few weeks down the line. I think this is because we don't have a lot of natural downtime any more and even when we physically stop doing things, our minds are constantly churning. Sadly, we have arrived at a stage in our evolution where we have to create a time artificially for our minds to be still and calm.

Emotional regulation in the moment

When you're experiencing an intense emotion or are in a state of stress and feel you are tipping out of your window, you will need to regulate your emotional state to bring yourself back to your equilibrium. First, you need to notice and accept what you are feeling in your body, paying attention to your heart rate, breathing, muscle tension and chemical sensations. Then calm the physiological reaction in your body using your breath to soften and relax muscle tension and slow your heart rate. The more attention you pay to your inner state on a regular basis, the less likely you will be to overreact to your children. Don't be tempted to skip this because it is this inner state that your children sense and your emotions won't just go away because you ignore them; they will rise to a crescendo whether you like it or not.

Remember, emotions are like waves that rise and fall; you don't need to be afraid of them or run away from them as long as you know how to regulate them. Third, and this works best when you are within your window and have already calmed the emotion at a bodily level, you need to activate those higher, rational parts of your brain by identifying and labelling your emotions, challenging negative or irrational thoughts that may contribute to them, and reframing the situation so you can adopt a wiser, more balanced perspective. Be aware of when you are tipping out of your window of tolerance as your children will inevitably follow. Once you are feeling calmer, you might decide to take some action such as speaking to someone about how you feel, or changing some aspect of the trigger situation. This is what we often rush to when we feel uncomfortable but in my view it is best to first soothe emotions, optimise thinking and only then decide what needs to be done, if anything.

Exercise your social engagement system so you can deeply connect with your children

Connecting with your child can occur through a state of excited love or quiet love. Children need both types of connection from us but the quiet kind of love can be hugely calming and restorative, especially for younger children and older children who are sad or upset. This involves becoming still, softening your body and opening your mind. Slow down the pace of your words, lower your volume and soften your tone of voice. Soften your face and upper body, especially the area around your heart, throat and chest. Look into their eyes and see the person there. You understand how powerful the social engagement system is and how we are designed to detect and respond to psychological safety. Prioritise this when you are around your children and notice the difference it makes.

When they are dysregulated, don't rush to speak; a shared and gentle silence, where you allow them to feel their feelings fully in the warmth of your support, can convey more meaning than a plethora of words, however eloquent. Don't rush to brush away their emotions because you are afraid of where they might lead. If they are receptive, touch them gently and lovingly, as if they are precious and delicate. Let them feel they are deeply cherished. It is this quality of 'presence' that nurtures them, their sense of self, and their ability to form close and meaningful relationships. It is the sense of emotional safety that comes from this state that really transforms them.

When mothers and fathers are under pressure, stressed, busy, depressed or otherwise struggle with self-regulation, their social engagement systems will switch off and reciprocity is compromised. Without being judgemental and with the compassion that every parent deserves in the face of their difficulties, we must think about how we support and help parents and children who face these challenges. Perhaps it's time to question the effects that our 'drive and strive'-fuelled, independent ways of living have on those children whose parents struggle with emotional regulation, particularly those who have to cope with raising children without adequate emotional support. Perhaps we need to protect our children against the relentless onslaught of technology that might reduce or eliminate the need for face-to-face interaction in the future. Given what we know about the social engagement system, emotional safety and human wellbeing, we need to protect these right-brain processes in an increasingly left-dominant world.

Develop a greater sense of compassion and loving-kindness for yourself and your children

This is one of the most powerful ways to transform your relationship with yourself and your children, and not only that, it has numerous health benefits too. Make the time to practise loving-kindness meditation and extend warm wishes towards yourself, your children and other people in your life. Even if you can't dedicate time to the full practice, just bring an image of their smiling faces into your mind, soften, relax and wish them happiness. The release of oxytocin that accompanies this is enormously calming and restorative and will help you to remain connected and empathic towards your children.

It is okay for you to struggle a bit with this because these feelings come more easily to some than to others. Don't be hard on yourself if you don't feel your heart warmed when you wish your children well. But keep at it, because most of us can develop and generate these feelings with practice over time. Once you become familiar with these feelings, try to incorporate them into your day-to-day interactions with your children. Compassion is a calm, happy feeling. It is a desire to alleviate the suffering of another person but it is different to being triggered by that person's emotions. Try to differentiate yourself and remain connected yet distinct from your children in your respective emotional experiences.

Let's not forget that children are often at the mercy of the bottom-up system

The prefrontal cortex and top-down system are in development until the mid-twenties, so our children are often controlled by their amygdaloid gut responses to things, without the ability to step back, reflect and reason with themselves. They are at risk of frequent emotional hijacks because they have a raging bottom-up system without the balancing power of the top-down system. This does not mean they are incapable of reflection or reason, much to the contrary, but that once they are in the grip of a reasonably strong emotion, they are more likely to succumb to it. They simply don't possess the capacity for self-regulation, self-awareness and moral reasoning that we might, for our own convenience, like them to demonstrate.

When children are in the grip of powerful emotions, we, as parents, need to step back and calm our own amygdala-based emotional reactions by depersonalising the situation and understanding that our children don't have the full apparatus needed to soothe their emotions. In terms of emotional regulation, children often use methods that we find difficult to tolerate, such as screaming, crying, arguing, ignoring and so on, to try to regain some sense of emotional equilibrium. When they do so, they are simply using the best technique they know of to even the scales in an attempt to feel safe. It is a sign that they need coregulation and help in learning better coping strategies, not a call for criticism, punishment or judgement. They can only do what their brains allow them to do at any point in time. Do encourage and help them to learn to manage their behaviour over time but please don't forget that it stems from their neural circuitry, most of which will have been shaped by their genetic material and their early experiences, neither of which was of their making.

Pay attention to sleep and diet because the mind, brain and body are interconnected

I can't begin to emphasise how important sleep is to emotional regulation. When we are sleep-deprived we cannot bring ourselves back to our emotional baseline and our ability to connect and remain regulated will be compromised. Children are even more vulnerable to sleep disruption and if there is one area where you need to be a little more rigid, let it be this one. Personally, I have noticed that once I have had a few late nights in a row, relaxation and empathy become a struggle and I become increasingly dysregulated. Start to tune in to when you are tired and at the earliest sign of it, let yourself have what your body is trying to tell you is needed.

Diet also plays a part in emotional wellbeing. We now know that the gut has over 100 million neurons, which has led some researchers to label it 'the second brain'. The gut produces several neurotransmitters, including a large percentage of our levels of serotonin, which is associated with positive mood. We can't ignore the fact that being busy, combined with our low tolerance of inconvenience and discomfort, has changed what and how we eat. Consuming high levels of sugar, packaged foods and foods that are low in nutrients will affect our emotional wellbeing, hormonal balance, mood and emotions. And just as importantly, because it is the slowing, calming lower vagal circuit that is responsible for digestion, eating quickly and on the go, without lingering long enough to notice, savour and appreciate our meals, may create more trouble in our relationship with food, our bodies and gut health.

PARENTING CHILDREN FOR BALANCE AND RESILIENCE
Try to accept your children as they are and set aside judgement

Abandon your mental blueprint for who you want them to be and give them the gift of truly seeing who they are. Accept what you see and cherish it, for this is what we all deeply crave – a sense of being accepted and understood as we are. Allow them to feel safe with you – a felt, 'in the body' kind of emotional safety, rooted in trust. A sense of trust born out of the knowledge that you are on their side, that you 'get' them and their feelings; that you gently hold their emotions, their vulnerabilities, their confusion and shortcomings in your heart without hardening towards them. If they can't have this from you, they probably won't seek it or receive it from anyone else because they will never see themselves as being worthy of it.

Accept them as they are and let them feel this through your way of being with them; remember that their view of themselves and how lovable they are, and their template for future relationships, is sculpted by their interactions with you. It is as precious as if they were to take their own hearts out and place them in your hands, fragile and delicate and vulnerable. This is not a burden but an honour. If you can strip away thoughts, rules and anxieties about how your child should be, or how their future selves might be, or even how comfortable your own life as a parent should be, and allow yourself to be fully present with your child, moment to moment, you will discover how enriching your connection with your children can be.

Try to challenge all the non-conscious demands you make of yourself, your children and the world around you. There is no law of the universe that states that we or our children should be tidy, accomplished, kind, responsible, perfect, punctual and so on. These are left-brain-generated rules, that when implemented with rigidity, will create layers of judgement, stress, negativity and resentment on both sides. For example, emphasising kindness in a left-brain way, i.e. through talking and telling rather than showing via the right brain, might only generate left-brain, rule-based kindness in your children. At some point, they may be capable of unkindness, even if it is in the interests of being kind. They may become aggressive and hurtful towards one person when they are defending the rights of another person whom they perceive has been treated unkindly by that person. You may prefer that your children act in certain ways because this brings you comfort, certainty, control or vicarious self-esteem but they are unique human beings with their own values, needs, personalities and desires and they will feel connected with you when you respect their individuality. By all means parent them in ways that encourage them to develop and live by the values that matter to you, whatever they may be, but don't expect your children to conform rigidly with them.

Let them reveal themselves to you without the fear of disconnection. Disconnection happens when you hear what they say, or see what they do, from a place of correction, self-focus, judgement, or anxiety. To connect deeply with your child means to soften yourself towards them – your heart, your mind, your rules, your insecurities. Allow them to show you their uncensored selves without the fear of being evaluated or shamed. But if you see that they are hurting themselves or others, or that they are struggling to find their way with something, guide them gently to a better outcome, without undue anxiety about what this might mean now or in the future.

I know you feel concerned about them; maybe parenting feels like a huge responsibility. But try to regulate that concern, because your anxieties will ultimately become their anxieties. Let them have the freedom to evolve in time, free from the pressure to develop in the best way possible, or in the fastest way possible. Give them the ability to accept themselves as they are, but to know that they can change and develop through reflection and persistent effort; that our brains are malleable and can be sculpted. This is very hard to do in practice but being emotionally regulated and cultivating compassion and mindful presence will build the foundation.

Connect before you correct

As all parents know, we can't survive our years together with our kids simply through connection, however lovely that ideal is! We also need to pay attention to 'correction' because good parenting is based on a combination of high warmth and high authority. Being connected and compassionate does not mean that we shy away from tough conversations with our children. Children need to learn to tolerate some distress, including being corrected when they make mistakes and understanding the consequences of their behaviour, however difficult that might feel. We must allow them to be pushed to the edges of their windows of tolerance but as parents we want to try not to push them too far out.

When we correct from a base that doesn't include warmth and connection, a child can easily feel shamed or threatened. When a child feels mild shame, he may still be able to comprehend what you are saying and it may act as a social deterrent for the behaviour that caused it, but when he feels a very high level of shame, for example, when someone he cares about shows disgust and contempt at him or his behaviour, the feeling of 'badness' is so threatening to his sense of self that it activates the threat and defence response in the brain and leads to a shutting-down of reason and sense-making.

Correcting your child requires a sense of perspective. Your job is not to make them perfect so that they, and you, can avoid emotional discomfort, rejection, failure or disappointment; so that they can be approved of by all and be guaranteed a special or safe place in life. Your job is not to make everything comfortable or enjoyable for them either. Difficult emotions are a part of life and inevitable for them, no matter how many parenting articles you read and how hard you try to give them all the right or best experiences. Your job is to be beside them, regulated and compassionate, as they make mistakes, experience discomfort and find their place in the world.

Doing this builds the connections in their brains that will allow them to soothe and regulate their emotions; it builds resilience. Allowing them to feel their feelings in the safety of your own support will embolden them; they'll realise emotions are just sensations that come and go. It will teach them that they needn't be afraid of vulnerability, of failure, of criticism or discomfort, because they will learn that no matter how uncomfortable it may feel at the time, they can feel okay again so there is nothing to be afraid of. But this only works if you can stay emotionally regulated and open yourself, even in the face of their distress, opposition or loss of control.

To do this, you need to regulate yourself into the calm and connect mode via your breath, voice and face, and let this kindness infuse your interaction with your child. When you have a well-functioning social engagement system, this becomes easier to accomplish. At the same time, whilst keeping your voice gentle but firm, you need to keep your language clear and direct, giving your child an unambiguous description of the behaviour you don't like, the consequences of it in terms of your feelings and the feelings of others, and what you want the child to do differently. Make sure you are making gentle eye contact so you can monitor your child's emotional response to what you are saying and adapt your message accordingly.

Try to discuss difficult issues by referring to the behaviour in question but not by labelling your child as a whole person. So, rather than saying, 'You're so messy! Just tidy up as I've told you!', you might say, in a slow voice, 'When you don't tidy up your mess, it isn't very fair on the rest of us who live here and like our home to be tidy. I don't mind you being messy while you play but at the end of the day, I expect you to clear up.' Remember, although what you say to your children and how you say it matters, it is still of secondary importance to how you both feel during the interaction because it is these invisible processes that shape and define the outcome.

Balance high warmth with high authority

Encourage them to do the right thing by setting high expectations for them, and communicate these standards clearly, directly and regularly. Believe in them and hold them accountable but take the time to prepare and support them consistently to meet your expectations. Don't give in to their demands when they put up a resistance, but when you say 'no', do it with kindness, with an empathic recognition that doing the right thing doesn't always feel good. And remember to be flexible sometimes about your expectations of them, just as you would want someone to be with you. After all, there are times when we all deserve a break. This teaches them adaptability and empathy. Expect a high standard of behaviour, give them the tools they need to be able to meet your expectations, but be understanding and kind when they fail because most of the time their failures are not an intentional act of malice but an inability to successfully self-regulate, the architecture for which develops slowly over time and is environment-dependent.

Children are capable of taking responsibility for themselves at a fairly young age. Where it is feasible, and with appropriate boundaries, show them how to do things for themselves, give them access to the things they need to take responsibility for themselves (for example, have the breakfast things at a low level so they can get their own breakfast) and be prepared for them to learn the consequences of their own actions and decisions. Ask them questions to ensure they fully understand the consequences of their choices, but once they have made the choice, let them experience those consequences for themselves, even if they are negative, because this is how they will learn and develop their own decision-making capabilities. Naturally this only works in scenarios where the consequences are likely to evoke a mild to moderate emotional response.

Homework, for example, in our home, is not something I get involved in unless they ask for my help. I have taught them that it is their responsibility to do their homework and if they don't want to do it, they must be prepared to explain their reasoning to their teachers at school. I ask them how they will feel when they have to speak to their teachers to explain why they haven't done it and whether they are prepared to feel that way. If they are struggling with something, they can ask me for help and I will always make time to do so. But I won't cajole, bicker or force them to do things because this makes us resent each other and takes us into threat and defence mode. With my younger child, whose memory is not quite strong enough to support full responsibility, I will remind her to do the things that she is responsible for. If I suspect they might forget to do things, I help them with developing their capability to remember but I won't do it for them. So, every morning, I will ask them, 'What do you have and what do you need?', but I won't pack their snack boxes or put folders in their bags.

Find a balance between being authoritarian and permissive.

As parents we need to be comfortable wielding authority but keep it balanced with warmth, kindness and respect for our children as individuals. For example, rather than controlling what they eat all the time, you might give them a choice between two or three things that you don't mind them consuming. This type of parenting approach builds and develops those higher cortical brain regions that play a part in decision-making and moral behaviour. Giving them choices and boundaries empowers them, respects them as people with thoughts and feelings of their own, distinct to yours, and helps foster connection and warmth.

Encourage your children to think independently by asking questions that help them comprehend things for themselves. Talk to them about how you, they and others feel but don't expect that they 'should' understand all this. They will understand when their brains develop the connections that enable them to grasp these things, possibly not until much later in life. This is okay. You don't have to make them perfect – just lay the foundations and give them the opportunity to do the rest in their own way and in their own time. They will not turn into failures if they don't accept all you want to offer them. Be gracious when they decline. They are not a reflection of you; they are individual people with their own needs, personalities and ambitions. Like us all, they are complex, fallible and imperfect but always deserving of your love and acceptance.

Connecting with your children does not involve praising them

A last note about warmth and authority: being loving, kind and warm does not mean praising our children. Praising them is similar to judging them; it encourages them to see themselves as objects in the eyes of others. Rather than praising and judging, try to focus on how things feel to them so they can start to understand their values, personalities, likes, needs, insecurities and drivers from within rather than extraneously. Show appreciation rather than praise. Let them develop a strong inner sense of themselves as unique, complex and fallible human beings who are too multifaceted to be rated and judged by themselves or anyone around them. This will help them to develop authenticity, self-awareness and self-respect, and over time will guide them towards life decisions that respect their emotions, needs and preferences. If you like to praise and find it satisfying, praise how much effort they put into things and the strategies they use, rather than them as a whole. For example, you might say, 'I like the way you took the time to colour that in so carefully' rather than 'You're so talented at colouring – wow!'

If you're interested in knowing more about this, Carol Dweck has done some wonderful work on how children develop theories about their intelligence and characteristics, and how these theories, which they can't articulate but which influence them all the same, determine how hard they persist in the face of challenges and difficulties. The mindset a child adopts can shape how they deal with failure, how stable their sense of confidence is, and how they ultimately perform on difficult tasks.

Final thoughts . . .

Developing children who are resilient and regulated is not a matter of 'doing' but a way of 'being'. I know this is not easy. Start with yourself. Learn to practise self-acceptance, self-compassion and mindfulness so that you can come to really 'feel' the message here, not just think about it, but live and breathe it. If there is only one thing you take away from this book, it is to prioritise connection above all the other bits of advice. And finally, remember that you are human too, with your own ups and downs, and you will not always be loving and open to them. As long as you are open to them more often than you are closed, and can build up a 'bank of trust' over time, you'll see that they become forgiving of your lapses and foibles. You'll find you get caught up in a virtuous circle, not a destructive one. When they feel deeply understood, accepted and cherished, they will want to listen to you. They will want to please you because it feels good for them to do so, and this will feed your connection and compassion for them. And when this happens, you, and they, will flourish together. And in time, when they extend this way of being into their own relationships with others, and into the efforts they expend, you will know there is no better gift you could have given them.

Acknowledgements

I cannot have written this book without the indelible influence of my father who sadly passed away in 1997 when I had just started university. My father encouraged me to question, to reflect and challenge the status quo though he never pressured me to be or do anything other than what felt right to me at the time. My mother, who raised my brother and me after the death of our biological mother when I was three years old, uncomplainingly put aside her needs and desires so that I could be educated in England, for which I am deeply grateful. I might not have arrived at this juncture in my life if not for my cousin, Piyush, who, through his generosity of spirit and kindness, rescued me when I was in danger of losing my way after my father died. His belief in my ability to overcome the odds and the pride he took in my achievements sustained me at a time when I might have crumbled. There are some things you can never repay.

My deepest gratitude goes to my husband who effortlessly embodies many of the values and characteristics I write about in this book. He is my rock, and in the twenty years I have known him he has provided a level of unfailing stability, kindness and decency that I value immensely. I am convinced that loving him and being loved by him changed my brain, my mind and my sense of self for the better in an enduring way. His ability to put my needs and the needs of the children consistently above his own is testament to the parenting style of my wonderful mother-in-law, whose love for us is the very definition of generosity – pure, unselfish and unceasing.

In helping me write this book, I thank my husband for his encouragement and benevolence in the face of my last-minute requests for advice and support, often in the middle of the night! I thank my children for so graciously understanding that writing this book has meant there have been times when I haven't been able to give them the attention they would have liked. I want them to know that I always put them, their need to spend time with me and their need to feel loved, first. My son, with his cheerful and resilient personality, had to remind me often that it was all going to be worth it, that I should put in the hard work, even at the weekends, and that they were happy to keep themselves entertained whilst I worked. I am grateful, throughout the process of writing this book, for the presence of my daughter, whose empathy, kindness and tolerance inspired and nourished me. They cared for me whilst I cared for them.

I want to thank a few people who have helped me with feedback and input along the way. My friends, Jo and Daniel, who read what passed off as early chapters, and my sister-in-law, Pri, who not only read the very first version of the opening chapter but who has patiently and supportively listened to two years of updates on the evolution of the book. My colleague, Rosa, who has been on hand to cast her wonderfully right-brain eye over some of the more creative aspects of the book. To all my friends, my family and my colleagues at VWA who have supported, encouraged and tolerated me during this long and intensive process, I want to extend my heartfelt gratitude and appreciation.

I am very thankful to Tom Asker at Little, Brown who has helped shape this book with his remarkable patience, insight and guidance. I thank Amanda Keats and Una McGovern for refining the text and making it presentable. I would also like to thank Giles Lewis and Nikki Read at Robinson How To who encouraged me to write a book on parenting when I had, initially, only thought about writing for the corporate sector. I cannot think of a more worthy topic to spend my time on than understanding how children are deeply shaped by parental love and I am appreciative of having had the opportunity to do so.

This book would not exist without the wonderful research it is based on – research that has opened my eyes and heart to the power of love and connection to alter our brains, minds, physical health and ways of experiencing the world. There are a few Titans in the academic world to whom I feel indebted and whose incredible work I can say, with absolute certainty, has changed my life and that of my clients and children for the better. In particular, I owe my deepest appreciation to Iain McGilchrist, Allan Schore, Stephen Porges, Daniel Siegel, Paul Gilbert and Barbara Fredrickson. Their work brings us full circle.

Bibliography

Baars, B. J. and N. M. Gage, *Cognition, brain, and consciousness: introduction to cognitive neuroscience.* 2nd ed. 2010, London: Academic.

Baumeister, R. F., et al., 'Bad is stronger than good.' *Review of General Psychology,* 2001. 5: pp. 323–70.

Baumeister, R. F., 'Self-regulation, ego depletion, and inhibition.' *Neuropsychologia,* 2014. 65: pp. 313–19.

Carter, C., S and D. W. Pfaff, *Hormones, brain, and behavior.* 2nd ed. 2009, Amsterdam; London: Academic Press.

Damasio, A. R., *The feeling of what happens: body and emotion in the making of consciousness.* 2000, London: W. Heinemann.

Davidson, R. J. and S. Begley, *The emotional life of your brain: how its unique patterns affect the way you think, feel, and live – and how you can change them.* 2012, London: Hodder & Stoughton.

Ekman, P., *Emotions revealed: recognizing faces and feelings to improve communication and emotional life.* 2nd ed. 2007, New York: Owl Books.

Fredrickson, B., *Love 2.0: how our supreme emotion affects everything we think, do, feel, and become.* 2013, New York: Hudson Street Press.

Fredrickson, B. L., et al., 'Open hearts build lives: positive emotions, induced through loving-kindness meditation, build consequential personal resources.' *Journal of Personality and Social Psychology*, 2008. 95(5): p. 1045–62.

Gilbert, P. J. and Choden, *Mindful compassion*. 2013, London: Constable & Robinson Ltd.

Goleman, D., *Emotional intelligence*. 1995, New York: Bantam Books.

Goleman, D., *Working with emotional intelligence*. 1998, London: Bloomsbury.

Green, J. A., P. G. Whitney and M. Potegal, 'Screaming, yelling, whining, and crying: categorical and intensity differences in vocal expressions of anger and sadness in children's tantrums.' *Emotion*, 2011. 11(5): pp. 1124–33.

Hanson, R. M., *Buddha's brain*. 2009, Oakland, Calif: New Harbinger Publications.

Heinrichs, M., et al., 'Social support and oxytocin interact to suppress cortisol and subjective responses to psychosocial stress.' *Biological Psychiatry*, 2003. 54(12): pp. 1389–98.

Hill, D., *Affect regulation theory: a clinical model*. 1st ed. 2015, New York: W. W. Norton & Company.

Hughes, D. A. and J. F. Baylin, *Brain-based parenting: the neuroscience of caregiving for healthy attachment*. 2012, New York: W. W. Norton & Company.

Jang, K. L., W. J. Livesley and P. A. Vernon, 'Heritability of the big five personality dimensions and their facets: a twin study.' *Journal of Personality*, 1996. 64(3): pp. 577–91.

Kok, B. E. and B. L. Fredrickson, 'Upward spirals of the heart: autonomic flexibility, as indexed by vagal tone, reciprocally and prospectively predicts positive emotions and social connectedness.' *Biological Psychology*, 2010. 85(3): pp. 432–6.

McEwen, B. S., 'Central effects of stress hormones in health and disease:

Understanding the protective and damaging effects of stress and stress mediators.' *European Journal of Pharmacology*, 2008. 583(2–3): pp. 174–85.

McGilchrist, I., *The master and his emissary: the divided brain and the making of the Western world*. 2010, New Haven, Conn.; London: Yale University Press.

McKay, M., M. Davis, and P. Fanning, *Thoughts & feelings: taking control of your moods & your life*. 3rd ed. 2007, Oakland, Calif.: New Harbinger; Enfield: Publishers Group UK [distributor].

Moffitt, T. E., et al., 'A gradient of childhood self-control predicts health, wealth, and public safety.' *Proceedings of the National Academy of Sciences of the USA*, 2011. 108(7): pp. 2693–8.

Norman, G. J., et al., 'Oxytocin increases autonomic cardiac control: moderation by loneliness.' *Biological Psychology*, 2011. 86(3): pp. 174–80.

Ogden, P., et al., *Sensorimotor psychotherapy: interventions for trauma and attachment*. 1st ed. 2015, New York; London: W. W. Norton & Company.

Okon-Singer, H., et al., 'The neurobiology of emotion-cognition interactions: fundamental questions and strategies for future research.' *Frontiers of Human Neuroscience*, 2015. 9: p. 58.

Porges, S. W., *The polyvagal theory: neurophysiological foundations of emotions, attachment, communication, and self-regulation*. 1st ed. 2011, New York; London: W. W. Norton & Company.

Raefsky, S. M. and M. P. Mattson, 'Adaptive responses of neuronal mitochondria to bioenergetic challenges: Roles in neuroplasticity and disease resistance.' *Free Radical Biology and Medicine*, 2017. 102: pp. 203–16.

Schore, A. N., *Affect regulation and the origin of the self: the neurobiology of emotional development*. 1994, Hillsdale, NJ: Lawrence Erlbaum Associates, Inc.

Schore, A. N., *Right brain psychotherapy (Norton Series on Interpersonal Neurobiology)*. 2019, New York: W. W. Norton & Company.

Segerstrom, S. C. and G. E. Miller, 'Psychological stress and the human immune system: a meta-analytic study of 30 years of inquiry.' *Psychological Bulletin*, 2004. 130(4): pp. 601–30.

Shahrestani, S., A. H. Kemp and A. J. Guastella, 'The impact of a single administration of intranasal oxytocin on the recognition of basic emotions in humans: a meta-analysis.' *Neuropsychopharmacology*, 2013. 38(10): pp. 1929–36.

Sieff, D. F., *Understanding and healing emotional trauma: conversations with pioneering clinicians and researchers*. 2015, Hillsdale, NJ: Routledge.

Siegel, D. J., *The developing mind: how relationships and the brain interact to shape who we are*. 2nd ed. 2012, New York: Guilford Press.

Siegel, D. J., *Pocket guide to interpersonal neurobiology: an integrative handbook of the mind*. 1st ed. 2012, New York: W. W. Norton & Company.

Siegel, D. J. and M. F. Solomon, *Healing moments in psychotherapy*. 2013, New York: W. W. Norton & Company.

Tronick, E., et al., 'The infant's response to entrapment between contradictory messages in face-to-face interaction.' *Journal of the American Academy of Child and Adolescent Psychiatry*, 1978. 17(1): pp. 1–13.

Weinberg, M. K., et al., 'A still-face paradigm for young children: 2(1/2) year-olds' reactions to maternal unavailability during the still-face.' *Journal of Developmental Processes*, 2008. 3(1): pp. 4–22.

Wigton, R., et al., 'Neurophysiological effects of acute oxytocin administration: systematic review and meta-analysis of placebo-controlled imaging studies.' *Journal of Psychiatry and Neuroscience*, 2015. 40(1): pp. E1–22.

References

1   Fredrickson, B., *Love 2.0: how our supreme emotion affects everything we think, do, feel, and become.* 2013, New York: Hudson Street Press.

2   Goleman, D., *Emotional intelligence.* 1995, New York: Bantam Books.

3   Goleman, D., *Working with emotional intelligence.* 1998, London: Bloomsbury.

4   Moffitt, T. E., et al., 'A gradient of childhood self-control predicts health, wealth, and public safety.' *Proceedings of the National Academy of Sciences of the USA,* 2011. 108(7): pp. 2693–8.

5   Baumeister, R. F., 'Self-regulation, ego depletion, and inhibition.' *Neuropsychologia,* 2014. 65: pp. 313–9.

6   Siegel, D. J., *Pocket guide to interpersonal neurobiology: an integrative handbook of the mind.* 2012, New York: W. W. Norton & Company.

7   Jang, K. L., W. J. Livesley, and P. A. Vernon, 'Heritability of the big five personality dimensions and their facets: a twin study.' *Journal of Personality,* 1996. 64(3): pp. 577–91.

8   Siegel, D. J., *The developing mind: how relationships and the brain interact to shape who we are.* 2nd ed. 2012, New York: Guilford Press.

9   Raefsky, S. M. and M. P. Mattson, 'Adaptive responses of neuronal

mitochondria to bioenergetic challenges: Roles in neuroplasticity and disease resistance.' *Free Radical Biology and Medicine*, 2017. 102: pp. 203–16.

10 Hanson, R. M., *Buddha's brain*. 2009, New Harbinger Publications.

11 Siegel, D. J., *Pocket guide to interpersonal neurobiology*.

12 Baars, B. J. and N. M. Gage, *Cognition, brain, and consciousness: introduction to cognitive neuroscience*. 2nd ed. 2010, London: Academic.

13 McGilchrist, I., *The master and his emissary: the divided brain and the making of the Western world*. 2010, New Haven, Conn.; London: Yale University Press.

14 McGilchrist, I., 'Hemisphere Differences and Their Relevance to Psychotherapy' in D.J Siegel and M. Solomon (eds), *Healing Moments in Psychotherapy*. 2013, New York: W.W. Norton & Company.

15 McGilchrist, I., *The Divided Brain and The Search for Meaning*.

16 Davidson, R. J. and S. Begley, *The emotional life of your brain: how its unique patterns affect the way you think, feel, and live – and how you can change them*. 2012, London: Hodder & Stoughton.

17 McGilchrist, I., *The master and his emissary*.

18 Baumeister, R. F., et al., 'Bad is stronger than good.' *Review of General Psychology*, 2001. 5: pp. 323–70.

19 Schore, A. N., *Affect regulation and the origin of the self: the neurobiology of emotional development*. 1994, Lawrence Erlbaum Associates, Inc.

20 Schore, A. N., *Right brain psychotherapy (Norton Series on Interpersonal Neurobiology)*. 2019, New York: W. W. Norton & Company.

21 Schore, A, and Sieff, D.F in 'On the same wave-length: how our

emotional brain is shaped by human relationships' in D. J. Sieff (ed), *Understanding and healing emotional trauma: conversations with pioneering clinicians and researchers*. 2015, East Sussex: Routledge.

22 Schore, A. N., *Right brain psychotherapy*.

23 Schore, A.N., *The Development of the Unconscious Mind*. 2019, New York: W.W. Norton & Company.

24 Hughes, D. A. and J. F. Baylin, *Brain-based parenting: the neuroscience of caregiving for healthy attachment*. 2012, New York: W. W. Norton & Company.

25 Damasio, A. R., *The feeling of what happens: body and emotion in the making of consciousness*. 2000, London: W. Heinemann.

26 Okon-Singer, H., et al., 'The neurobiology of emotion-cognition interactions: fundamental questions and strategies for future research.' *Frontiers in Human Neuroscience*, 2015. 9: p. 58.

27 Ekman, P., *Emotions revealed: recognizing faces and feelings to improve communication and emotional life*. 2nd ed. 2007, New York: Owl Books.

28 Green, J. A., P. G. Whitney, and M. Potegal, 'Screaming, yelling, whining, and crying: categorical and intensity differences in vocal expressions of anger and sadness in children's tantrums.' *Emotion*, 2011. 11(5): pp. 1124–33.

29 Hill, D., *Affect regulation theory: a clinical model*. 1st ed. 2015, New York: W. W. Norton & Company.

30 Ibid.

31 Ogden, P., et al., *Sensorimotor psychotherapy: interventions for trauma and attachment*. 1st ed. 2015, New York; London: W. W. Norton & Company.

32 Siegel, D. J., *The developing mind*.

33 Schore, A. N., *Affect regulation and the origin of the self.*

34 Schore, A, and Sieff, D.F in 'On the same wave-length: how our
    emotional brain is shaped by human relationships' in D. J. Sieff (ed),
    *Understanding and healing emotional trauma: conversations with
    pioneering clinicians and researchers.* 2015, East Sussex: Routledge.

35 Ibid.

36 Schore, A. N., *Affect regulation and the origin of the self.*

37 Baron-Cohen, S., *Zero degrees of empathy: a new theory of human
    cruelty.* 2011, London: Allen Lane Publishing.

38 Porges, S. W., *The polyvagal theory: neurophysiological foundations
    of emotions, attachment, communication, and self-regulation.* 1st ed.
    2011, New York; London: W. W. Norton & Company.

39 Baumeister, R. F., et al., 'Bad is stronger than good.'

40 Schore, A. N., 'Early interpersonal neurobiological assessment
    of attachment and autistic spectrum disorders.' *Frontiers in
    psychology*, 2014, 5, 1049. doi:10.3389/fpsyg.2014.01049

41 Hill, D., *Affect regulation theory.*

42 Tronick, E., et al., 'The infant's response to entrapment between
    contradictory messages in face-to-face interaction.' *Journal of the
    American Academy of Child and Adolescent Psychiatry*, 1978. 17(1):
    pp. 1–13.

43 Weinberg, M. K., et al., 'A still-face paradigm for young children:
    2(1/2) year-olds' reactions to maternal unavailability during the
    still-face.' *Journal of Developmental Processes*, 2008. 3(1):
    pp. 4–22.

44 Baumeister, R. F., et al., 'Bad is stronger than good'.

45 Carter, C., S and D. W. Pfaff, *Hormones, brain, and behavior.* 2nd ed.
    2009, Amsterdam; London: Academic Press.

46 Wigton, R., et al., 'Neurophysiological effects of acute oxytocin administration: systematic review and meta-analysis of placebo-controlled imaging studies.' *Journal of Psychiatry and Neuroscience*, 2015. 40(1): pp. E1–22.

47 Shahrestani, S., A. H. Kemp and A. J. Guastella, 'The impact of a single administration of intranasal oxytocin on the recognition of basic emotions in humans: a meta-analysis.' *Neuropsychopharmacology*, 2013. 38(10): pp. 1929–36.

48 Heinrichs, M., et al., 'Social support and oxytocin interact to suppress cortisol and subjective responses to psychosocial stress.' *Biological Psychiatry*, 2003. 54(12): pp. 1389–98.

49 Norman, G. J., et al., 'Oxytocin increases autonomic cardiac control: moderation by loneliness.' *Biological Psychology*, 2011. 86(3): pp. 174–80.

50 Carter, C., S and D. W. Pfaff, *Hormones, brain, and behavior.*

51 Kok, B. E. and B. L. Fredrickson, 'Upward spirals of the heart: autonomic flexibility, as indexed by vagal tone, reciprocally and prospectively predicts positive emotions and social connectedness.' *Biological Psychology*, 2010. 85(3): pp. 432–6.

52 Segerstrom, S. C. and G. E. Miller, 'Psychological stress and the human immune system: a meta-analytic study of 30 years of inquiry.' *Psychological Bulletin*, 2004. 130(4): pp. 601–30.

53 Fredrickson, B., *Love 2.0.*

54 Fredrickson, B. L., et al., 'Open hearts build lives: positive emotions, induced through loving-kindness meditation, build consequential personal resources.' *Journal of Personality and Social Psychology*, 2008. 95(5): p. 1045–62.

55 Gilbert, P. J. and Choden, *Mindful compassion.* 2013, London: Constable & Robinson Ltd.

56 Hanson, R.M., *Buddha's brain*.

57 Burns, D., The Feeling Good Handbook. 1999, New York: Plume

58 Fennell, M.J.V. 'Depression', in K. Hawton, P. Salkovis, J. Kirk and D. Clark (eds), *Cognitive Behaviour Therapy for Psychiatric Problems: a Practical Guide*. 1989, Oxford: Oxford University Press.

Index

Note: page numbers in *italics* refer to diagrams.

IBS *see* irritable bowel syndrome

immune system 2, 205, 218

impulse control 9, 10, 11–12, 21, 24, 43, 61, 69, 168–9

independence
    encouragement 243–5
    premature 126–7, 143–4

individualism 195

inflammation 179, 205, 218

inner state, paying attention to your 235

insula 95, 160, 180, 206

intelligence *see* Emotional Intelligence

Intelligence Quotient (IQ) 9

internal representations 128–9, 130

interoception 95–6, 159

Interpersonal Neurobiology xvii, 16

intimacy 70, 143, 175, 194–5

introversion 171

intuition 26, 42, 44, 55, 56, 86, 159–60
    see also gut feelings

IQ *see* Intelligence Quotient

irritable bowel syndrome (IBS) 131, 182

job stress 215

joyfulness xvi, 18, 62, 70, 107, 144, 193–4, 199

judgement 147–9, 201
    judging children through the eyes of others 6

parenting without 58, 59–63, 201, 202, 236, 239–41, 245
and self-acceptance 62–3
and self-image 62–3

kindness 14, 59, 69, 112, 150–3, 186, 233, 240, 242–4, 247–8
    see also loving-kindness; self-kindness

labelling 31, 40–1, 45, 53, 64, 235, 242–3
    positive 63
    see also mislabelling

left hemisphere 36–52
and busyness 216
characteristics 45–6
and compensation 104
and connection 55–76
differences from the right 39–46
and disconnect 232
dominance 48–50, 54, 56–8, 60–1, 64, 85–6, 212, 216
and the drive and strive state 212
and emotional processing 85–6, 108
and emotional regulation 110–11
grasping nature 40–2, 50, 55
and how you see the world 39–40
and parenting 49